Illustration of Celtic mythic imagery from the Gundestrup Cauldron.
The central figure may be Dagda who is associated with cauldrons.

THE SECRET ACADEMY

WE CAN'T GIVE IT AWAY

www.secretacademy.com

A CIP catalogue record for this book is available from the
British Library

ISBN13: 978-1-906069-11-7

Typeset by Wooden Books Ltd, Glastonbury, Somerset.
Printed and bound in the UK

the
SQUEEZE
PRESS

ANCIENT METROLOGY

VOL. I: A NUMERICAL CODE

Metrological Continuity in Neolithic, Bronze, and Iron Age Europe

John Neal

Besides this, they have many discussions as touching the stars and their movement, the size of the universe and of the earth, the order of nature, the strength and the powers of the immortal gods, and hand down their lore to the young men.

-

Julius Caesar concerning British Druids,
from *Commentarii de Bello Gallico*

CONTENTS

PREAMBLE
May, 2016

THESE VOLUMES were originally conceived as a single volume of collected works that were to be presented in the temporal order in which they were written. In the course of time and with inevitable delays, the collection of articles grew to a size that precluded them from being presented altogether. They are therefore split into three volumes of linked topics.

The three articles following these preambles concern megalithic rings and brochs, these occupy the same landscape but are separated by millennia in the eras of their construction. It was noticed quite early on in the study, that clusters of objects of a similar nature are quite the best method with which to apply comparative metrology. In this way certain constructional formulae that were universally used, particularly relative to the pi ratios and square roots, are most obvious.

It was decided to include the article – on what ought to be – the unrelated Greek stadia lengths of the surviving running tracks in this volume. This is because there can be no doubts raised as to what their constituent units of measurement actually are – with exactitude; the start and finish lines are still in place upon the ground and have been competently measured. In this way, some of the modules that were used by the ancient Greeks are clearly identified, and are found repetitively in the ancient British monuments to high degrees of accuracy.

The findings concerning the identification of modules of measurement and the applications of these modules in the field, owe a great deal to the discoveries of the, recently deceased, philosopher-antiquarian, John Michell. He emphasised the importance of "number" inasmuch that in the monuments and descriptions of divine cities logical and canonical number was being expressed, and these essential numbers can only be incorporated through the modules of measurement.

Such correlating of canonical number, the written word, metrology and philosophy, synthesizing with cosmological and geodetic schemes and constants into a singular system, may be concisely expressed; a word for this universal synthesis was forwarded by Dr. Hans Ulrich Vogel of the University of Tubingen, it is "Metrosophy."

Dr. Vogel is a Sinologist and used the term to describe this fusion of these seemingly disparate disciplines in Han China as:

"Metrosophy" may be defined as 'number speculation within cosmological philosophemes', but from a more inclusive perspective this is certainly too limited a definition, because also the relationships between magical, religious and political thought on the one hand and metrology on the other have to be taken into account. Both in metrology and metrosophy numbers are of great importance. In metrology, numbers are used for defining and counting measure and weight units, while in metrosophy they serve as basic stuff for the creation of systems of number symbolism, magical numbers, and correlative numerology."

Further: *"There is no doubt that we are dealing with a typical product of Han correlative thinking which endeavored to systematize, categorize and integrate phenomena of cosmos, nature, and human realm with the help of yin-yang, Five Phases and other numerological concepts as well as by adopting systematic textual techniques."*

Such insights markedly encapsulate those of John Michell who had similar views of metrology to Vogel; views that he had arrived at quite independently and had largely deduced through the works of Plato, Pythagorean lore and scripture. The advantage that John possessed in the decipherment of metrology was that he thought in terms of the English foot as his base module; had he thought in terms of the metre the subject could not have been understood.

This point will be made many times in the ensuing articles; this is not a reactionary stance nor partisanship, it is a simple and true observation. The metre, without any kind of justification, is regarded with virtual reverence by those who regard themselves as "rational" and "scientific." Yet, there is no hard and fast definition of the metre in terms of the English foot or any other of the foot measures of antiquity.

In Britain, the metre was not defined with fair precision until 1895 by Jean-René Benoit and Henry James Chaney as 3.28084275ft and this was the value previously used by Col Alexander Ross Clarke in his geodetic surveys from 1858-80. This is a foot of .3047997m and this was *rounded* to be exactly .3048m in America. There was a similar rounding by the Japanese to make their *shaku* (foot) equal exactly .303m; and so it was with all nations, the metre was merely *close to* their native measures. When the Smithsonian geodetic tables were published as late as 1897 they carried the caveat that they were to be expressed in English feet because *"in order to convert them into metres it is necessary to adopt a ratio of the foot to the metre, inasmuch a precise determination of this ratio is now in progress...."* this was one hundred and six years after it was devised.

This indistinct quality of the metre in relation to any other system of mensuration, is entirely due to the hasty adoption of the provisional metre as the national standard by the revolutionary government of France in 1799, known as the *mètre des Archives*. Although its inaccuracies and failings were pointed out to the investigating committee at the completion of the final geodetic survey of the Paris meridian, the provisional metre was considered too troublesome to correct. There is something aesthetically very unsatisfactory about this arrangement. It has left us with a universal "scientific" standard only fit for the bankers and the counting house, because all it can do is quantify, and we are not entirely sure of its length against any established other length.

A true measurement cannot stand alone; it must be a physical length in comparison to another or others; the unit fraction integration of ancient systems allow for total accuracy and eternal preservation by such comparisons. True to the tenets of "metrosophy," by using traditional measurement one is invariably dealing in whole and rational numbers in whatever order of natural phenomena that one is attempting to classify. For example, the distance to the sun from the earth, our astronomical unit, is exactly 100 million common Egyptian 5,000 feet miles. The International Astronomical Union defines the distance to be 149,597,870,700 metres; which would you prefer?

This is not to imply that the ancient Egyptians were aware of this fact it merely illustrates the point that with, what is essentially a simple method, one may select the most appropriate module to rationalise any distance into easily remembered units. This point is made in the appendices to the Greek Stadia Lengths (see p. 244-6)

During the recent years of rampant colonialism throughout the world, it was customary, in port towns and cities, to fire a noonday cannon from the citadel for the convenience of the general populace and the ships at anchor. This custom is still observed in Hong Kong and Cape Town. There is an apocryphal story of a curious tourist on one of the minor Caribbean islands who enquired of the town gunner how he kept such accurate time to fire the cannon at noon. The man replied that he had a watch that had been made by the craftsman Swiss clockmaker, whose shop dominated the town square; in the window of the shop was the finest and most accurate clock conceivable, and he kept his watch calibrated by this impeccable instrument.

The curious tourist, when he was next in the town square, enquired of the clockmaker how he set his clock with such accuracy. He was duly informed that the clockmaker calibrated his master clock to the noonday cannon.

This is a self-reinforcing myth as a form of tautology that is exactly analogous to the nature of the metre as an accurate instrument of universal reference. Since all pretense of the metre being the equivalence of one ten millionth of the earth's quadrant arc has been ditched; after many recalibrations and definitions of the metre to criteria other than this inaccurate geodetic relationship. It is now regarded since 1983 by the National Institute of Science and Technology *as the length of the path travelled by light in vacuum during a time interval of 1/299 792 458 of a second.* The charming little addition to this is the rider: *"Note that the effect of this definition is to fix the speed of light in vacuum at exactly 299 792 458 m. s -1 ."* This must be the noonday metre.

Either one of these nonsensical definitions could have been brought into the realm of logic by having a reinforcing reference. The clockmaker, from time to time, could have made corrections by judicious use of a gnomon; and the National Institute of Science and Technology could have given another example, such as the Imperial yard, in the same speed of light definition. We would then know how long the metre is in comparison with the yard, because at present, *nobody* knows the length of the metre in comparison to any other module of antiquity with exactitude.

The problem of the length of the metre is symptomatic of a deep and fundamental problem with modern science as a whole. There is the oft-repeated cliché that science is the new religion, inasmuch that it is faith based and the catechism of scientific dogma is little more than a belief system. Each of the tenets, a sort of Ten Commandments, on close inspection prove to be too irregular to be "laws." This is because in a specific sense, each of the "constants" is anything but *constant*. Gravity and the speed of light, in particular have proven to be variable, at different times and at different locations.

In a less specific way, the general cosmology and cosmogony of the new religion is equally unempirical; because it relies so heavily on mere assumptions. The invention of dark matter and dark energy being an example, both of these hypothetical forces were invented to justify what appeared to be happening to matter in the universe; there was not sufficient matter to hold it together, and there was insufficient energy to contribute to its expansion. Similarly, the red shift of light gave rise to the belief that all matter was retreating from a central point at an increasing rate; by playing this phenomenon backwards, it was predicted that everything had emerged from a single point out of nothing, in one gigantic explosion.

The elusive metre is a deserved offspring of this barren soulless mechanism, a very spoiled only-child, unsuited to sacred architecture or "metrosophy." Had man, in prehistory, devised a single and arbitrary unit such as the metre in order to regulate their tribal and national affairs, we would have no architectural history; the Great Pyramid, Teotihuacan, Angkor Wat, the Forbidden City, Stonehenge or any other example of monumental or cultural significance, could have been neither designed nor constructed. Everything would be either purely utilitarian or a self-indulgent flight of decorative fantasy such as Dubai.

Ancient metrology bears no resemblance to the modern system in its concept, its modules or their application; for these reasons it is impossible to understand when expressed in terms of the decimal system, because the essential numbers are destroyed. It is most easily made clear when articulated in the British imperial units because it is a system as ancient as any other, directly related the majority of the world's measurement systems and indirectly related to all the rest.

In modern usage of measurement, numbers are of no importance; in architecture, the designer simply has the available space, which is then divided up into purely utilitarian modular divisions according to its purpose without regard to the number of modules. A modern building may be entirely composed of fractured numbers of the metre. Even in the repetitively found modular features, such as door and window heights and widths, stair risers or fixed furniture heights and depths, no numerical regularity would be discernible in module terms. Whereas in ancient or historical building the modules of construction were of paramount importance, and are relatively easy to deduce. For example, in many styles of architecture, Palladian for one, certain ratios are repetitively used in spatial design such as the double square, the three by four rectangle, five diagonal – or other Pythagorean triples. In which case, the modules would all be selected to express a whole module in the ratio *numbers*. The three, four, five rectangle, for example, would be of three, four and five – yards, paces, fathoms, decempeda, pertica, or even larger modules according to the scale of the building. Number would be of primary significance, almost instinctively so on the part of the designer. This aspect of design has been completely obliterated by the introduction of the singular SI measurement.

The modern attempts to fix the length of the metre to the wavelength of light could never lead to a completely rational system because it is reliant for accuracy upon two variables – the speed of light and the "second." The speed of light varied between measurements taken in 1928 and 1945 by 20 kilometres per second (Sheldrake), and the second is constantly lengthening due to the slowing of the earth's rotation. The metre will therefore be true to its criteria only during a certain period in history (albeit the totality of human history). This may seem to be hair-splitting, because we know that a dull workable utilitarian system has been devised based upon this shifting sand – but due to its temporary nature, it is somehow illusory. The speed of light is determined by the *frequency* with which a *wavelength* of light passes a fixed point and since Ole Roemer's experiments, in 1673 there has been an ongoing quest to determine what this is.

Reported in New Scientist, in October of 1983, concerning the 17th General Conference on Weights and Measures, it was announced that the speed of light would no longer be subject to revision and *"Any refinement in the accuracy of measurement would only alter the length of the metre. (? ed) The hunt for the speed of light will be over!"* This too, appears to the layman as another obvious fudge. At a meeting of the Royal Society in January 2011, 150 participating scientists set no less than seven constants governing all mensuration possibilities with great finality, stating that each of them "*were exactly....*" That's it then, we don't have to worry anymore; until the next series of redefinitions, that is.

Consequently is there anything at all, in a universe subject to constant change, that may be considered inarguably permanent upon which a true, eternal and natural measurement may be based without recourse to quantum mechanics? According to Pythagorean doctrine, the one unchanging thing that governed all phenomena was number itself, and number is the ultimate reality. Similar to the modern search for permanence in wavelengths, ancient societies such as the Pythagoreans saw the numerical significance in sound waves. This was the division of a string into the 4th 4:3 the division of the string at the 1/4 mark, the 5th, 3:2 the division of the string at the 1/3 mark and the octave 2:1, the division of the string at the ½ mark.

The resemblance to ancient metrology of this arrangement is that unit fractions provide the connective links; all that is lacking is the physical length that may be termed base one. In ancient China, the original measurement system was also based upon sound, the pitch pipe or *huangzhong*, and there *was* a physical length attributed to it. Unfortunately, nobody remembered for sure what it was, and this resulted in a protracted search for the original length of the perfect pitch by the court musicians throughout all of the Dynasties. This went on until modern times since the time of the Yellow Emperor of the 3rd millennium BC – who designed the original. This rather makes our 200 years of infighting over the length of the metre look like a brief enquiry.

The nature of mensuration, from the most ancient of times until the present, is most clearly evidenced in the course of this collected works to be a singular system that is based upon pure number. It is integrated by unit fractions of the different foot modules, and the deliberate variations within each of these clearly defined modules are also governed by lesser unit fractions; but exactly how the physical length of base one – which to all intents and purposes appears to be the English foot, was determined – remains unanswered. Because this fundamental requirement to a "wrap" on the subject has proved elusive to the present author, does not mean that the question is unanswerable, and comfort is taken from the knowledge that neither the Pythagoreans or the Chinese – nor anybody else, has forwarded a watertight answer to this basic question.

DEDICATION

I GREW UP for the most part in rural surroundings and cannot remember when I first took an interest in methods of measurement and the modules that were used at that time. When such things as the "Cheshire mile" or corn reckoned as so many "quarters" to the acre and so forth, were mentioned I always felt a surge of interest. Even back then in the nineteen forties and fifties such things having passed from common usage were barely remembered; only the old folk in the smoky public bar could remember fragments of them, not the teachers in our schools.

I studied agriculture and always had a keen interest in archaeology and history, particularly British history, and realized that the units of measure in use at that time had an incredibly long pedigree and were an intrinsic part of our cultural heritage as a nation. Just how long this pedigree is was brought home to me by reading Maud Cunnington's account of her excavations at Woodhenge, which she so named as being contemporary with nearby Stonehenge. The massive sets of concentric ellipses that governed the posthole geometry she deduced were laid out in increments of a measurement unit of 11½ inches (Thom's later accurate survey of Woodhenge proved her right because it is better interpreted in terms of a Roman foot of exactly 11.52 inches rather than his "megalithic yard," I was later able to calculate that she was correct to one part in 575). This is explained in the recent addendum to "Measuring the Megaliths." (*see p. 146*).

I was intrigued by this information in as much I had believed that the Neolithic and Bronze age people were too primitive to possess or even need a sophisticated system of measures. Ever after, this simple fact gave me an enormous respect for my remote ancestors, utterly banished was my erstwhile opinion of cultural superiority to ancient people.

Although I was interested in these antique measures it never occurred to me to either study the subject in depth or even *contemplate* writing it up. Even then it was fast disappearing and this was many years before the acceptance of metrication as our primary measurement system was considered; now that this is a fait accompli it may yet be the death knell. I hope to make it clear that this would be a sad loss to civilisation.

It was not until I met John Michell when in my mid twenties that my latent obsession with metrology was awakened. At that time he was researching the material for *The View Over Atlantis* and we would often discuss the forthcoming content. It was not only me that was keenly interested, that book inspired a whole generation. For the first time people had a concept shattering glimpse of John's vision of the high culture that must have arisen in prehistoric times and had spread over the whole world. The evidence for its existence was scattered across the globe preserved in the temples, monuments and megaliths. They were carefully sited and their interrelationships often took the form of alignments that would have to have been methodically surveyed.

The evidence was not confined to the monuments; the same cultural influence had brought about the monotheistic scriptures, both early Christian and Hebrew. Through the Gnostic science of Gematria and Jewish Cabbala much more than the literal word was conveyed in them. This is because in both the Greek and Hebrew alphabets the letters are also their numbers, resulting in words, phrases and even whole sentences having a numerical value and these numbers that emerged were often of a canonical nature. That is, the repetitive values that occur such as 432, 864, 1080 etc. can be and were applied to many classes of phenomena. In all religions, Hindu, Greek Pantheism, Hebrew, Norse and Celtic to name some of them, these numbers repeat in their mythologies and foundations of their Heavenly cities, in their time cycles and orderly arrangement of the universe.

These same numbers that emerge from myth and scripture are also applied to the measurement of the heavens and most interestingly to the Earth and its principal luminaries. Expressed in English miles the sun is 864,000 miles in diameter and the moon is 2,160, were the earth spherical it would have a diameter of 7,920 miles. The plans of the heavenly cities as described to Ezekiel and to St John by a divine messenger holding a measuring reed of six sacred cubits were described in these canonical number values.

John went on to demonstrate how a number series applied to monument, in terms of the measuring units, revealed that many monuments were as a fractal with the dimensions of earth. This was very heady material and it prompted us both to explore the field of metrology in greater depth and we began to realize that the modules of measure that he was using at the Great Pyramid and methods of multiplication to a geographic level were seriously in error. This in no way collapsed the overall theory but we knew that the matter of metrology had to be refined.

It was about then that I moved to America for around four years and we lost regular contact. We were both working on the solutions unaware that the other was similarly engaged. My own research seemed to indicate that the basic foot measures and other modules had an affinity with cycles of time, for example if the English foot represented the solar year then the Roman foot was proportionally the length of lunar year and the remen of 20 Roman digits (the palimpes) would be 444 days. I was building a whole cosmology centered on cycles of time translating into linear measure. Worse, it seemed that these time cycles were immortalized in the ancient monuments, the best example of which was the dimensions of the Great Pyramid. A basic ancient Egyptian module was a length of 5 royal cubits, surveyors cords were knotted at 5 cubit intervals, this is one 56th part of the pyramid height. Inspired by the *The View Over Atlantis* data I continued the series division of the monument by a factor of 56. 56 cubed is 175,616 and the pyramid height divided by this number is a miniscule length (roughly a 30th of an inch) and 365.2422 of them exactly equalled the English foot.

This reasoning led to the conclusion that the intended height of the Great Pyramid was 480.82067ft and the length of the royal cubit would correspondingly be 1.7172166ft. The problem is, that all of these measures that I had found were acceptably close to the actual values and nobody could gainsay me for the simple reason that at that time no module from the ancient world had ever been defined as an absolute. Everybody involved with metrology had always had to express the values between plus and minus parameters. I therefore continued in my beliefs and researches that ultimately proved to be far from the facts.

I had not the slightest problem in dropping my inventions, that had taken years to refine, when confronted with the truth of the matter, for the simple reason that my years of investigation had qualified me to recognize that truth. Measurement by its very nature must be axiomatic; measurement is the determination of any magnitude in reference to a fixed magnitude that is regarded as a standard. Time cannot be that fixed reference because however slight it may be, it is a variable.

ANCIENT
METROLOGY

*The Dimensions of Stonehenge
and of the Whole World as
therein symbolised*

JOHN MICHELL

It was in 1978 that I returned to England and, struck up my friendship with John where we left off, and we discussed metrology at length. He was unconvinced by my reasoning as to the foundations of the measures and although he had nothing better to offer at that time, he was working on it and this was palpably obvious. I had to return to my affairs in Wales so we saw each other but rarely. It was in 1981 that he walked through my door like the cat who got the cream, bastard, and gave me a copy of *Ancient Metrology, The Dimensions of Stonehenge and of the Whole World as Therein Symbolized.*

He had not exactly beaten me to it; I had not a hope in hell of ever finding it. All I have ever been able to do is to confirm the evidence that he presented, I have merely elaborated upon his simple foundation. His small book was the most succinct distillation of the subject; he was the first man ever to have found the answers to the questions that the great and good had laboured over. For these reasons, these collected works are dedicated to his name.

The following tract, written by John Michell in 1973 as the first of a series of pamphlets in his "Radical Traditionalist" series that addressed all manner of social issues in his inimitable style, it encapsulates the facts concerning ancient metrology, denigrates the adoption of the metre and gives the reasons.

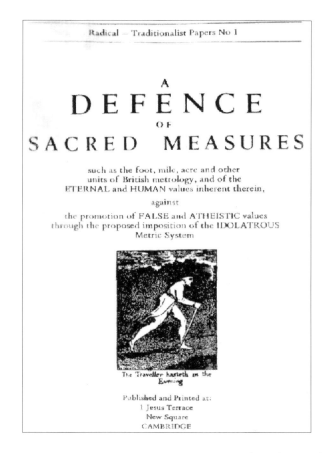

A DEFENCE OF SACRED MEASURES by John Michell

"1. It is necessary, for the reasons set out in this pamphlet, to oppose with the utmost vehemence the proposal to abandon our traditional standards of weight and measure in favour of the metric units invented at the French Revolution.

2. The fundamental reason for emphasising the importance of this issue is that it clearly epitomises the conflict between two contrasting philosophic or cosmological points of view. The first is expressed in the ancient belief, orthodox for many thousands of years and only recently denied, that eternal and human values (the two being inseparable) should govern the social order and institutions. According to this view, human nature, though always varying, remains essentially the same in all times. There is one philosophy, one religion, one science, though for each of these there are infinite expressions. Man and cosmos are alike in this, that both are constant and made in the same image. Both are therefore measured by the same standards, and for this purpose the ancient units of metrology were designed. Against this view is the newly received idea that the universe and its inhabitants are ever subject to purposeless change mechanically operated; that eternal values are a delusion, human values incidental; that the only true religion is idolatry, the worship of the chance material products of the time; and that the interests of the state idol are more important that those of the human individual and should be consulted above all else. By this reckoning the body of the idol provides a more appropriate source and standard of measures than the human frame, and this is the origin of the metric system.

The point at issue is whether an enlightened, humane and scientific civilization should adopt as its standards of measure units such as those still in use in England, which were formerly, and for good reason, regarded as sacred, having the advantages of tradition, inherent meaning and natural application, or whether the metric system, which has none of these qualities nor any of its own to replace them, is the more appropriate.

3. According to the first of the philosophies contrasted in the preceding paragraph the adoption of sacred standards of measure, together with an appreciation of their significance and value, is essential in order to achieve and maintain true civilization.

4. Sacred measures are those units which relate to natural constants on more than one scale and demonstrate the unity between the macrocosmic body of the universe and the human microcosm. The present British units, the foot, mile, acre etc, are by this definition sacred; the metric units are

not. The origins of the two systems and the implications in their use are contrasted in the following paragraphs.

5. The foot and the other linear and land measuring units that relate to it are of indefinable antiquity. They were known to the Sumerians, Chaldeans and the ancient Egyptians and appear once to have been universal, for they survive in different parts of the world, wherever the interests of the people are still given precedence over those of modern technology and commerce. Their advantages for all human purposes are obvious. A carpenter gauges an inch by the width of his thumb and its tenth part by his practiced eye; a builder estimates the length of a wall by the two yard span of his outstretched arms, and a surveyor paces by the yard. Cloth is sold by the cubit, the distance from elbow to finger tip, and other such units as the span and handbreadth were formerly used which have now generally become obsolete. Of course no two people have the same bodily dimensions, and the canonical man has never existed save as an idea or archetype. These traditional units are not, however, imprecise or inaccurate. Ancient societies regarded their standards of measure as their most sacred possessions and they have been preserved with extreme accuracy from the earliest times. A craftsman soon learns to what extent the parts of his own body deviate from the conventional standard and adjusts accordingly.

Sacred units of measure apply not only to the human scale but also to the astronomical. For this reason they were said, at a time when such language was more generally understood, to have been 'revealed' to men, not invented by them. The purpose of ancient science was to maintain and invigorate as esoteric tradition, the primeval heritage, rather than to pursue innovations, not, as evolutionists have supposed, because of any deficiencies in the positive intellect of early men, but because education was formally directed on Platonic principles towards the development of the inherent sense of proportion by means of musical and mathematical studies, with the result that cosmology, the science of discerning and codifying reality, was respected above all. According to Plato in *The Laws*, the stability of ancient civilization was maintained by the application of a canonized law of proportion, a code of musical harmonies, to which artists and musicians were obliged to refer in all compositions. The canon was essentially numerical, capable of being interpreted in the appropriate terms for use in the various arts and sciences, as music, architecture and astronomy, and extending to such matters as theology and the art of government. Its corresponding geometrical expression was the figure, conceived as the synthesis of all geometrical types, which St. John described as the ground plan of the New Jerusalem and Plato as the mystical city of The Republic and The Laws. This figure was the symbol of the cosmos, and its dimensions, measured by the sacred units, the most important being the English foot and mile, reproduced the principal dimensions of the solar system, revealing accurate knowledge in some remote age of the measurements of earth, sun and moon.

6. It is thus claimed on behalf of units of measurement such as the foot, furlong, mile etc, whose preservation has hitherto been the honourable charge of the British nation, that they have a profounder significance than as mere arbitrary standards of length; that they are integral in the human view of the universe and cannot therefore be excluded from any social scheme founded on human rather than idolatrous principles. The philosophy, which provides the justification for their use, recognizes the existence of a natural law, reigning within both human nature and the universal soul, some knowledge of which is essential to the orderly conduct of human affairs. The word human is here emphasised, because the interests of the true science and of the people are not naturally opposed but complementary, and when the principles of this science are again established, as they inevitably must if the reenactment of the destruction of Babylon on a more grandiose scale is to be averted, the conduct of affairs will be directed towards the benefit of the people as a whole rather than of one class, the financiers and industrialists. In this event, the advantage of adopting sacred units of measurement, those which are inherent in the natural order and not simply the reflection of a transient, atheistic political philosophy, will again become apparent.

7. The history of the metric system, by which it is proposed to replace our traditional system of metrology, is indicative of its character. At the time of the French Revolution, when the Goddess of Reason was ceremonially installed in Notre Dame, a number of people, many of whom

were by all other standards apparently sane, were struck by the remarkable notion that the facts of nature, even the cycles of the sun and moon as manifest in the weeks and months of the calendar, might be varied by government decree. The revolutionary calendar, with its 10 month year and 20 hour day, the most spectacular feat of idolatry since the Tower of Babel, collapsed at once, but its companion, the metric system, was successfully imposed on the French people. According to Napoleon who lightened the penal sanctions by which its use was enforced, *"it violently broke up the customs and habits of the people as might have been done by some Greek or Tartar tyrant."* Despite popular rejection and following a number of bloody riots in which opposition to the compulsory use of the metre was suppressed, the metric system survived in France and was extended in the interest of uniformity to other European nations, always with the active assistance of the police or military.

8. The metre was originally intended, following ancient precedent, to be a geodetic or earth-measuring unit, one ten-millionth part of a quadrant of the meridian measured through Paris. Its length was finally established in 1798, as accurately as the scientific methods of the time allowed, as equal to 39.37 inches. A particular reason why this length so commended itself to its inventors was that it corresponded to no existing or traditional unit. In other words, it was purposely designed to be unlike any unit which had ever been found convenient in actual use. The old sacred measures, properly understood, promote harmony, stability and knowledge. The new atheistic system, conceived in ignorance and arrogance and nurtured on the blood of the people, is the fitting servant of the forces of greed and materialism that are currently favouring its adoption in England.

Note. The French might have acquired a true sacred and scientific system at the beginning of the 18th century had they adopted Carrini's proposed geodetic foot, equal to one six- thousandth part of a minute of arc on the terrestrial meridian or 1.013ft. This length is the same as the Greek foot by which the Parthenon was laid out and which, like all ancient units, was geodetic in reference.

9. This defence of the foot against the metre is based on two qualities that distinguish the foot from its rival. First, the foot is the established

measure of the British people and has been so from the earliest times, at least since the building of Stonehenge. It is universally known and is used in many countries including America, where on account of the republican common sense and practicality of the people it is to be retained. To abandon it to enforce the use of an alien system can in no possible way benefit the public interest. The compulsory introduction of the metre, which in the improbable event of popular opinion being consulted, certainly would be rejected by the great majority, is thus clearly defined as an act of tyranny.

The second argument in favour of the foot may to some appear excessively mystical. It is however the more essential of the two, and is here included for the consideration of those who are sufficiently experienced to understand its implications, and with an appeal for the indulgence of those who are not. The foot, as stated above, is a sacred unit of ancient cosmology, designed to illustrate the hermetic philosophy of "man as the measure of all things" and to promote harmony on earth by assisting the influences of true proportion to become active in human affairs. The way of thought that attends the use of the foot locates the centre of the world within each individual, and encourages him to arrange his kingdom after the best possible model, the cosmic order. The ancient method of acquiring this model was not astronomy but initiation, for those who presented themselves, suitably prepared, to the priests of Hermes were admitted to the study of the sacred canon, which demonstrated the link between the created, visible world and the creative world of archetypal notions, and provided the criterion for the discernment of truth and illusion. Those insufficiently curious to seek initiation could rest assured on the word of initiates, such as that given by Plato, that "things are far better looked after than we can possibly conceive".

10. That the inch, foot, furlong, mile, acre etc, are of very ancient and sacred origin has been demonstrated elsewhere. The fact may not appear of any great interest at the present time; but sacred means preordained and eternal, and to these epithets neither the metric system nor the theories behind its promotion have any claim. Naturally, each generation has the right to select whatever system of measurements it finds most appropriate, but it must then be content to be judged by its choice. It is therefore

the right and duty of those concerned, before the final decision is made between the foot and the metre, to consider carefully the origin, history and meaning of the two systems in order to see which one is most in accordance with their ideals and interests and best designed to promote civilized human values.

11. In Jung's phrase, the balance of the primordial world is upset. The art of government, as practiced from the earliest times, is to discern and weigh the various interests within the community, preserving dynamic stability by application of the mystical law of proportion. The decline of this esoteric science and the fragmentation of the canonical society prepared the way for the development of the new philosophy of civilization, which attaches more importance to external form than to essential reality. In consequence, the respect formerly given to the concept of a sacred order based on eternal values was transferred to the 19th century doctrine of "the survival of the fittest", a pernicious phrase, favoured by dictators, millionaires and other modern aberrations, and advanced by them in justification for all excesses.

The proposal to introduce the metric system into England is another episode in a lengthy historical process, by which the natural rights of the individual, including that of participation in decisions affecting his own immediate interests, have been eroded, often by measures ostensibly designed to protect them. The destruction of local independence that followed from the Reformation; the confiscation of church and common lands on behalf of the state and its monopolists; the extinction of the labourer's small-holding, compensated for by the benefits of the poor law and workhouse; by such events the process is illustrated. Throughout the 18th century the central government increased its power over the people whose interests it was shortly to betray by the following proceeding. Soon after the Napoleonic war, the government adopted a remarkable theory of economics, which held that the country's wealth could be increased by the simple expedient of printing more money. This new paper money was issued through bankers and jobbers, to whom it mostly adhered, thus generating a new dominant class, whose wealth and influence soon exceeded all others. This class is strictly parasitic, because it neither creates nor produces anything of human value, nor does it profess to rule for the benefit of the people as a whole. Its growing influence has led naturally to a corresponding decrease in the fortunes of everyone else. Its values have become universally accepted and embodied in theories of government, the natural function of rulers – to balance the various interests within the community – being disregarded. Finally, the native parasites have now been swallowed up by others larger and more anonymous, so that it has become no longer possible for the individual to identify the source of the authority by which he is governed.

The religion that has been engendered by this process is idolatry and the idol is that very "image of the beast" described by St John in Revelation 13, whose ritual is the worship of material form. The appropriate unit by which this idol is measured is the metric system: and so it is proposed. Yet, vast and inflated though it is, the idol is but a created thing with no claims to immortality. After the nature of such monsters its appetite ever increases, and each year ever greater sacrifices are demanded of the people, until the time comes for its destruction, and of this there is no lack of portents. In contrast, the foot belongs to a tradition of which it has always been said that, even though it may be suppressed and vanish for centuries, it will always recur, for its spores are deeply embedded in human nature, and the truth to which it refers is constant and unique. The submergence of this tradition coincides with the dark periods of history; with its cyclical rebirth the light of civilization is restored. To institutionalise the dark ages by giving authority to the metric system would be an act of folly inconceivable in any other age but our own."

Radical - Traditionalist Papers No. 1

BIBLIOGRAPHY

Proof of the ancient and sacred origin of the present units of British metrology is to be found in the following:

A.E. Berriman, *Historical Metrology*, Dent 1953.

John Michell, *City of Revelation*, Garnstone, 1972.

John Michell, "*The Sacred Origins of British Metrology in Britain: A Study in Patterns*, 1971, R.I.L.K.O., 36 College Court, W.6.

The bibliography quoted by John is indicitive of the minute data base from which he reached his correct conclusions regarding metrology. Apart fom Berriman there was very little on the subject that was accessible outside of the academic fraternity, it was before the advent of the internet and Stecchini at that time was unpublished.

In spite of John's aversion to the metre it is truly an ancient measurement in both length and origin. The problem with the metric system is not the physical length of the metre, it is simply how it is subdivided that utterly divorces it from ancient and natural standards. It has an exclusive decimal subdivision that precludes a rational subdivision by the number three. Yet this is what the metre essentially is – it is three "Belgic" feet in length. If the metre is so regarded, then in its relation to the earth it is the eminently suitable duodecimal 120,000,000 Belgic feet to the perimeter.

It is this metre that was basic to the ancient Greek estimates of the circumference of the earth as forwarded by Posidonius and Ptolemy. This metre length prompted Petrie to remark that if the British had adopted the Belgic foot as their standard, which they could easily have done because it was commonly used in Medieval Britain, then – "*But little adjustment would have been necessary for us to conform to the metre.*"

I may be forgiven for relapsing into allegory to describe the sequence of events, from the moment that John Michell saw the interconnectedness of the system to the present. He revealed the tip of an iceberg; I have manged to describe its totality, but the glacier from whence the iceberg calved, remains unexplored.

Book I

THE EXACT SCIENCE OF
ANCIENT
METROLOGY

"A true hereditary and traditional measure is something almost as vital as a language, sometimes more so; and to stamp it out utterly and instantly is not within the province, or the power, of any government of men for the time being."

- Charles Piazzi Smyth

John Michell, seated upon the steps of the Assembly Rooms,
Glastonbury, home of the "Megalithomania" conferences.
So called from his book of that name.

THE EXACT SCIENCE OF
ANCIENT METROLOGY

1.1 How John Michell First Established Certain Rules

John Michell (9 February 1933 – 24 April 2009) the great philosopher-antiquarian was the first man to establish definite values for units of measurement.   Prior to this, due to variations in linear measurement, virtually any close value could be proposed as an intended module. Through many comparisons of recorded modules he stated that examples were found in Roman, "Polar" feet, Greek, royal Egyptian and Ancient Jewish that all existed in two distinct variants that related as 175 parts to 176. This inferred that the modules of these different nations all referenced an earlier canon, which in turn implies an earlier and now forgotten culture that had been universal. These are his *exactly* expressed modules in English feet:

	TROPICAL	NORTHERN
Roman	.96768	.9732096
Polar	.987428	.993071
Greek	1.008	1.01376
Royal Egyptian	1.145454	1.152
Ancient Jewish	1.3824	1.390299

The two variants are under the headings "Tropical" and "Northern" because it was believed at that time that it was the variations in the length of the geographic meridian degree at 10° latitude and latitude 51° that dictated the lengths. These variants enabled him to calculate two specific earth radii from the "sacred" cubit of the Jews, that of 20,854,491ft for the polar radius and 20,901,888ft as the mean radius of the earth and that these radii differed as 440 to 441.

These exactly expressed values enabled the foundation of the more complex structure of ancient metrology to be understood because these data were now in place. Firstly it was realized that the fraction 175 to 176 was also the difference between the anciently used pi ratios of 25/8 (or 3.125) and 22/7 (or 3.142857), thus establishing a purely mathematical as well as the geographic explanation as to its presence in metrological values. Another focusing was the observation that the values of the "polar" feet at six parts to seven of the royal Egyptian, were in fact the universally known common Egyptian feet.

Also pertinent to the structure was the recognition that the term "sacred Jewish foot" at 1.3824ft was far too long to be termed "foot" – it was arrived at through the division of the sacred cubit by 1½, when in point of fact the sacred Jewish cubit is a two feet cubit. Two important factors were brought to the fore when the correction was made. Firstly, take the value 2.0736ft as the sacred cubit, this divided by two is 1.0368ft, it is therefore a two-feet cubit of the common Greek foot. Furthermore, if the value 1.3824ft is divided by 1½ it is .9216ft and this is a value of the *pie* or Spanish foot, one third of the vara; it is an Iberian 1½ feet cubit. Thus was revealed an exact integration of metrological values of far greater complexity than John had first proposed that goes much further than merely two variants, or this limited number of foot modules.

Just this small exercise is a perfect illustration of the total amalgamation of metrological values that underscores the point that it is a singular system. The sacred Jewish is common Greek and relates to Iberian as 9 to 8; the common Greek is 21 to 20 of the common Egyptian and 9 to 10 of the royal; the Iberian is 14 to 15 of the common Egyptian and 4 to 5 of the royal. In this fashion, all of the foot values of the world so interrelate– by unit fractions. Having knowledge of this connectedness and more than one value to maintain accuracy (by comparisons) it enables *any* module from the antique systems to be identified and *classified*.

1.2 THE ALGORITHM

Prior to the world's adoption of the Système International d'Unités, there was already an international system of measurement. It had become fragmented into what superficially appeared to be an unconnected collection of quite separate methods of mensuration. On close inspection of these, from all periods of history and from all quarters of the world, it is manifestly obvious that there was only one "system" of measurement. The greatest impediment to an understanding of this ancient science is the metric system itself.

No traditional unit that was replaced by the metre is expressible by an exact exchange; since it is so utterly divorced from the system that it supplanted no historic module of measurement is an exact number of millimetres.

Further to this, in the classification of modules, whatever the multiple lengths (stadia, miles, leagues etc.) that are encountered, they should first be reduced to their constituent foot length and the comparisons made at that level, but the metre has no subdivision that approximates to a foot. It is often difficult to identify any of the ancient multiple feet modules under consideration when they are expressed in metres because the prefixing numbers are lost. This will be become obvious in due course.

The algorithm or axiomatic rules of the structure of metrology, is that all of the foot modules of measurement are connected through a series of unit fractions, and all of these foot lengths have identical variations in their lengths. It is these variations that have occluded the detection of the whole-number interconnectedness of the various feet – because the wrong variants are most often compared, but the comparison must be made like for like.

The following list is of the core variants of the Greek feet; it is more extensive, but these are the variants most commonly encountered. The Root or number one is the English foot.

GREEK/ ENGLISH	*Root Reciprocal* 0.994318ft (30.307cm)	*Root* 1ft (30.479cm)	*Root Canonical* 1.005714ft (30.654cm)	*Root Geographic* 1.011461ft (30.829cm)
	Standard Reciprocal 0.996578ft (30.376cm)	*Standard* 1.002272ft (30.549cm)	*Standard Canonical* 1.008ft (30.724cm)	*Standard Geog.* 1.01376ft (30.899cm)

ROOT RECIPROCAL	175:176	ROOT	175:176	ROOT CANONICAL	175:176	ROOT GEOGRAPHIC
440 ⋮ 441		440 ⋮ 441		440 ⋮ 441		440 ⋮ 441
STANDARD RECIPROCAL	175:176	STANDARD	175:176	STANDARD CANONICAL	175:176	STANDARD GEOGRAPHIC

These are the fractions that govern these variations in the same feet.. The significance of these unit fractions, that of 441/440 and 176/175, is that they both have the property of yielding whole-number solutions in diameters and perimeters of circles in differing modules of measurement. If diameters are multiples of four the rationalisation of pi that was used historically is 3.125 as opposed to the more accurate 3.<u>142857</u> and 25/8 is 175 to 176 of 22/7. This means that a diameter of four Roman feet would have a perimeter of twelve Greek feet; the ratio of the Roman foot to the Greek is 24 to 25.

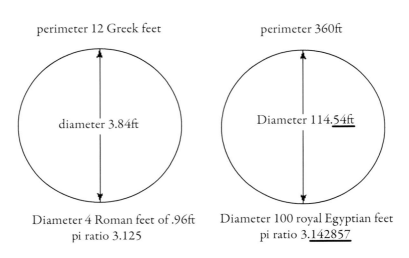

perimeter 12 Greek feet   perimeter 360ft

diameter 3.84ft   Diameter 114.<u>54ft</u>

Diameter 4 Roman feet of .96ft   Diameter 100 royal Egyptian feet
pi ratio 3.125   pi ratio 3.<u>142857</u>

It is the 440th fraction that is used on the diagram to the right; the royal Egyptian foot relates to the Greek as eight to seven, this is 1:1.<u>142857</u> but the addition 440th fraction increases the foot of the diameter to 1.<u>145</u>ft. This is how these fractions – both those relating the module lengths and those of the module variations – were utilised to give integer solutions in designs.

ROMAN	Root Reciprocal	Root	Root Canonical	Root Geographic
	0.954545	0.960000	0.965486	0.971003
	(29.094cm)	(29.261cm)	(29.428cm)	(29.600cm)
	Standard Reciprocal	Standard	Standard Canonical	Standard Geog.
	0.956715	0.962182	0.967680	0.9732096
	(29.227cm)	(29.394cm)	(29.562cm)	(29.731cm)

Listed above are the core (most commonly found) variations of the Roman feet; .96 *Root* differs from the Greek Root of 1ft as 24 to 25. It is because we are using the English/Greek foot to express these values that the Greek feet may also be regarded as the formulae by which all other foot lengths are classified. Other modules that also relate as 24 to 25; the Sumerian and royal Egyptian for example, would also share the same relationship of integer radius and perimeter.

There are nineteen *potential* Mathematical feet and they all evidence the identical variations. The term "mathematical foot" distinguishes the module from the "natural foot" – these natural or anatomical feet are half cubits. The list of the mathematical feet of *all nations* is shown below at their "Root" classification expressed in English feet. The list shows one of the unit fraction linkages between the modules in the offset column; these all relate through **square** numbers that are shown in bold type.

English ft	Fraction			Identification
.9	**49** 63			ASSYRIAN FOOT
.914285		**64** 80		IBERIAN FOOT
.925714			**81** 99	NOT IDENTIFIED 6:7 OF BELGIC
.935064			**100** 120	NOT IDENTIFIED 6:7 OF SUMERIAN
.942857			99 **121**	SAMIAN FOOT
.952381			80 **100**	LESSER ROMAN FOOT
.964285		63 **81**		NOT IDENTIFIED 6:7 OF NIPPUR
.979591	48 **64**			COMMON EGYPTIAN FOOT
1	35 **49**			ENGLISH / GREEK "OLYMPIC" FOOT
1.028571	36 48			COMMON GREEK FOOT
1.05	**49** 63			PERSIAN FOOT
1.066666		**64** 80		PERSEPOLITAN FOOT
1.08			**81** 99	GREATER BELGIC FOOT
1.090909			**100** 120	SUMERIAN FOOT
1.1			99 **121**	SAXON FOOT
1.111111			80 **100**	ARCHAIC ENGLISH FOOT
1.125		63 **81**		THE FOOT OF NIPPUR
1.142857	48 **64**			ROYAL EGYPTIAN FOOT
1.16666	**49**			RUSSIAN ½ ARSHIN

Three of the foot lengths are marked "unidentified" this does not mean that they are never encountered, but are too rarely found to give a positive nomination. They should be regarded as "potential" but must be there to give the connective unit fraction links; additionally, the canonical height of a man is six feet – and in one or another of the mathematical feet every man is six feet tall. The relationship of the anatomical foot to the height of a man (or woman) is of one seventh; these non-conforming feet would be legitimate on this basis alone.

The complexity of ancient metrology is further compounded by the fact that many of the Root values have a greater or lesser identification that vary by 1.008 (126/125), which is a combination of both fractions, – 176/175, plus 441/440. The most commonly encountered example of this reduction of the "Root" classification occurs in the Roman foot; there is no doubt that the unit faction linkage between the Greek and Roman feet is that of 25 to 24. The Greek feet also relate to the Egyptian foot as 7 to 8 but the Roman foot of .96ft has no unit fractional link with the Egyptian; the Roman foot must be reduced by its 1.008th part to equal .9523809ft and is then 5 to 6 of the Root royal Egyptian foot of 1.142857ft. At this reduced value it maintains its unit fraction link with the Greek but at the ratio 20 to 21.

Another module that often displays this property is the Belgic foot; it has been accepted since antiquity that the Belgic foot is 9 to 8 of the Roman. At a Root value of 1.08ft this displays no unit connection with the Greek, yet reduced by its 1.008th part to 1.071428ft it is 15 to 14 of the Greek. At this reduced value when encountered it is referred to as the "Doric" foot. This brings us on to the identification terminology that is met within metrological research.

1.3 IDENTIFICATION OF THE UNITS OF METROLOGY

It has been demonstrated that the term Doric is also a variant of what is termed Belgic. This practice of naming feet after the location, architectural style or nation in which they are found leads people to assume that there are a far greater number of quite different foot modules than actually exist. This is clearly evidenced by the spread of terminology that is applied to what is termed here as the "Assyrian," or the *least* mathematical foot at .9ft. It and its variants have been termed Lesser, Archaic, Italic, Oscan-Italic, Oscan-Umbrian, Campanian, neo Babylonian, Lydian, Assyrian, Mycenaean, Basque, Geometric – and many more.

The convention here is to collect all of the *variants* of the same basic foot into a common terminology by calling them after the variant that relates to the English/Greek foot by a unit fraction; in this case it is the Assyrian foot that is .9ft; then all of the variants as encountered in the above list would then be "Assyrian" followed by their classification term – *Root Geographic* for example (this example would be the so called Lydian foot which is .9 x 1.0114612). All foot lengths that are encountered may be precisely identified in a similar fashion. Generally speaking, a module may be classified by dividing the example in question by the closest *Root* value, but care must be exercised, for the following reasons.

Because of the closeness in value of certain of the listed foot lengths, in the spread of their differences – a greater variant of a lesser foot can sometimes exceed the length of a greater foot at its lesser variation. Caution must therefore be exercised in matters of identification; most often the intended module can be identified from the *context in which it is found*. One such example would be the foot termed Archaic English; it is the foot of the "yard-and full-hand," at 10 to 9 of the English, it is 1.111ft and at its greater variations has overlaps with the Nippur foot that is 9 to 8 of the English at 1.125ft.

The *Root Geographic* variant of the Archaic English foot is 1.12385ft and it is easy to see how this may be likened to the Nippur Root foot of 1.125ft. But the situation in which this variant was encountered was in calculating the foot of a geographic degree recorded in China as being 180 li in length. The convention is that the geographic degree contains 360,000 geographic feet. Therefore the variety of foot may be calculated by dividing 360,000 by, in this case, 180 and this will give the number of feet in the li; in this case it is 2,000 *Root* English feet. A li consists of either 1,500 or 1,800 "feet" so this must be an 1,800 feet *li* of 1.111 feet; then adjusted in length for the latitude in question (ca. 40°) it is 1.1239ft. This is how this particular module is identified, and illustrates the point that the Root foot that it resembles is not necessarily the correct interpretation. All other lengths, wherever they are encountered can be similarly classified, largely through the context of their use.

It has here been broached that in certain circumstances, module variations are governed by geographic considerations; certain of the variants accommodate the lengthening degrees of the earth's meridian in order to maintain the same *number* of feet to the degree at different latitudes. The geographic foot *variations* that are common to all of the different "systems" are 1.008, this is termed Standard Canonical, 1.011461, this is Root Geographic and 1.01376 is Standard Geographic. In terms of the Greek feet they are the lengths of the geographic foot at 10°, at 38° and at 51°. There are other localized variations that are met depending on the latitude, but the above variants are found universally and are the lynch pins of the geographic structure.

1.4 THE GEOMETRICAL NATURE OF "CUBITS"

As with all modules of measurement, the descriptive terminology often carries a wealth of interpretation; mile, league, stadia, furlong, iteru, parasang etc. cover a wide range of measured distance that are termed the same unit in different localities. So it is with the cubit; often enough it merely means the "covid" or "module." Halves or doubles of any measure often have the same term applied to them. Anywhere, from the Far East to Europe to the Americas, in ancient texts one finds references to the "step" of 2½ feet and this is sometimes termed cubit, half this length, termed *remen* or *palimpes* or *pygon* or *omitl* may also be referred to as cubit.

The basic cubit is of 1½ feet and is the length of the forearm and outstretched hand, the antebrachium or *ulna*, whence comes the term *ell* – is also a very general module term. The Aztec *cemmolicipitl* has the same meaning – one elbow. This is termed a *short cubit* and the *long cubit* is of two feet and these two are the true cubits. Short cubit, long cubit and step relate as 3 – 4 – 5 and are therefore geometrical ratios. It is through the basic geometry of the human form and the shape and dimension of the earth that the canon of measure stems.

The structure of ancient metrology was devised to express integers in designs. To this end, incommensurable ratios particularly *pi* and √2 are rationalised through close approximations. These two ratios are intimately interconnected through the following geometry.

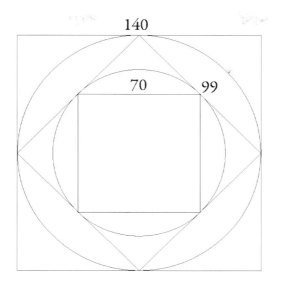

The commonly used approximations of √2 – 99/70 and 140/99

As illustrated above, the two fractions 99/70 and 140/99 are divisible by 7 as one factor and by 11 as the other. As both factors are common to the most often used pi ratio of 22/7 it is not surprising that these two approximations prove to be related through the medium of measurement – squares and circles and how they are reconciled as being basic to metrology. The most obvious relationship is that a square inscribed in a circle shares the diagonal of the square as the diameter of the circle.

99/70 equals 1.4142857 and this number exhibits the recurring 7th fraction that is constantly found in metrological values; 140/99 is 1.414141 and exhibits the recurring 11th fraction also commonly encountered in metrological values. These two approximations of √2 when multiplied together *are* exactly 2. Additionally, as so often happens, the numbers expressed in ratios are often measurements when expressed in English feet and the number/length, 1.4142857 is the Samian 1½ft cubit that stems from the foot of .9428571ft. Just as with the varying values of pi being selected to produce whole numbers in squares and circles, so it is with these values of the square root of two. It becomes obvious which value is being applied, because it will give an exact number of the modules in question in the solutions.

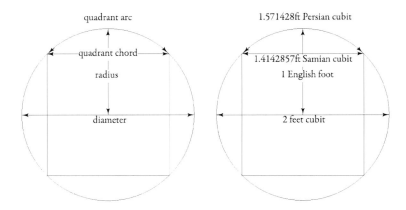

The basic module of all methods of mensuration is the foot. The basic factor that governs the dimension of a circle is the radius. The nature of the relationship between feet and cubits is exemplified in the nature of the circle and circumscribed square. If the radius is one foot (in any of the feet) the diameter is a two-feet cubit and the quadrant chord of the circumscribing circle is a one and a half feet cubit of a different foot. It is also true that most often, the quadrant arc is also a cubit module.

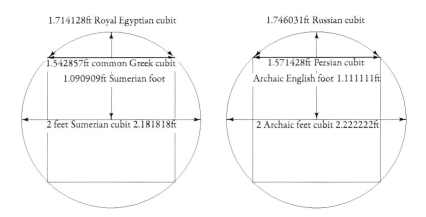

The top diagram to the right illustrates how the English/Greek foot radius propagates differing modules in the other dimensions; the Samian cubit in terms of the English foot *is* 99/70 as a *number*.

The next set of diagrams illustrate how integer modules arise from the same geometry, the resultant integers being maintained by values that differ from the Root classification radii by using the 176/175 or 441/440 fractions. Which of the exact approximations for √2 that are used (140/99 or 99/70) is dictated by the integer solution being a proven module. The feet that are used in the above radii are English feet and greater. If lesser feet are used, shown below, the quadrant chord may reduce to a 20 digit remen cubit.

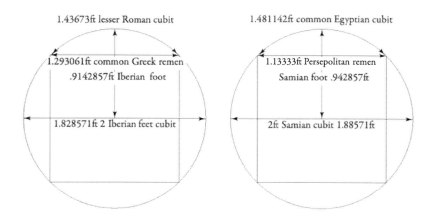

The diagrams above illustrate how these lesser feet, namely the Iberian and Samian radii, do not form 24 digits cubits but are 20 digits remen as the quadrant chord integer. It is fair to term a remen as a cubit because it is the forearm and clenched fist; the fist reduces the long palm length (10 digits) by one short palm (4 digits) thereby reducing the 24 digits true cubit to the 20 digits remen cubit. (This point will be made clear in the section on the human canon as the source of measurement.)

Another pertinent fact in the above geometry concerning the Samian foot radius is that the value one uses for pi, in order for the quadrant arc to be a common Egyptian cubit is that of 864/275, also known as the "Fibonacci pi" of 3.14<u>18</u> (this point will be clarified in the next section concerning the dimensions of the world as the secondary source of measurement.) Both exemplars, Man and the World, illustrate how the *ratio one uses*, either for pi or √2, must be selected to produce a proven module in the resolution. The best example of this point concerning the √2 ratios is the dimensions of the Great Pyramid as illustrated by the following diagrams.

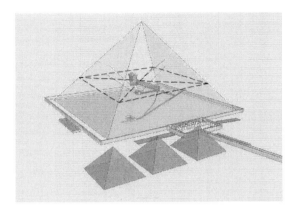

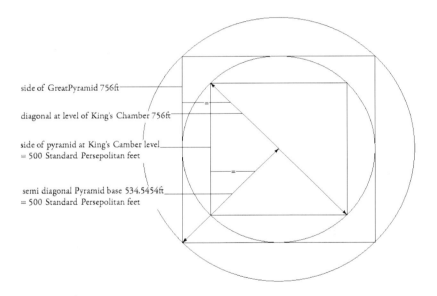

side of GreatPyramid 756ft

diagonal at level of King's Chamber 756ft

side of pyramid at King's Camber level = 500 Standard Persepolitan feet

semi diagonal Pyramid base 534.5454ft = 500 Standard Persepolitan feet

For many reasons, it is the south side of the Great Pyramid of 756 English feet that is the datum of the visible structure. The King's Chamber level has the ratio of 1 to √2 with the base of the pyramid. If the diagonal at that height would be the 756 feet of the base side, then the side of the reduced pyramid would be a five hundred feet stadium in terms of the Persepolitan foot, at 534.5454ft – if the √2 ratio used is 99/70. Then the diagonal of the pyramid base is twice that at 1,000 Persepolitan feet but the value for √2 that is used to achieve this precise number is that of the alternative – 140/99.

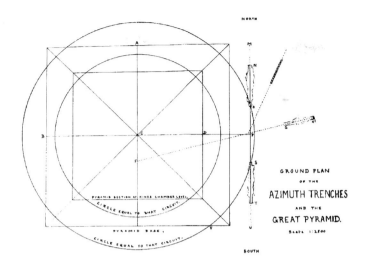

GROUND PLAN
OF THE
AZIMUTH TRENCHES
AND THE
GREAT PYRAMID.
Scale 1:2500

The diagram above is Flinders Petrie's rendition; it is not only illustrative of the √2 relationship regarding the base of the pyramid to the King's Chamber level, but it twice demonstrates the pi ratio as the quadrature of the circle, whereas the previous diagram depicting √2 uses the *inscribed* square. It is the most often quoted geometrical basis of the pyramid design, that of the height corresponding to the radius of a circle of which the base is the length of its circumference. In order for the geometrical functions of the Great Pyramid to be demonstrated, it is of paramount importance that the correct dimensions are used in calculation; these are height – 481.09<u>09</u>ft and datum base side 756ft, or 7 to 11. All modules that are used in the pyramid are of the *Standard* variation – *Root* plus the 440th part.

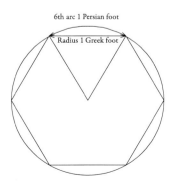

6th arc 1 Persian foot
Radius 1 Greek foot

Another example of how the pi ratio governs the modules of antiquity is the straightforward way that the radius of a circle relates to the 60° arc of the perimeter, rendering both as related modules. Once again, if this occurrence is shown to be regular throughout the range of feet, it is more clearly evidenced if the module of the radius is given the extra 440th part that transforms it to the "Standard" classification; it makes things a little clearer by the resultant arc being reduced to the "Root" classification. If the Root Greek foot (which is the English foot) as depicted in the diagram above, is extended by its 440th fraction to be 1.002272ft then the Persian foot of the arc is 1.05ft. This is then a Root value in the solution, this is not true across the board of all of the different feet; sometimes a Root radius will yield a Root arc as so: if the Root Assyrian of .9ft is the radius, the arc is the *Root* Samian of .942857ft.

It is through these comparisons of radii to the 60° of arc of the resultant circles that the manipulation of modules by the fractions that govern the module variants is most obvious:

Radius	*Arc*
Assyrian .9ft –	Samian .942857ft
Iberian .9163636ft –	Roman .96ft
Roman .96ft –	Greek 1.005714ft
Greek 1.008ft –	Persian 1.056ft
common Egyptian .981818ft –	common Greek 1.028571ft
common Greek 1.030909ft –	Belgic 1.08ft
Persian 1.05ft –	Saxon 1.1ft
Persepolitan 1.060606ft –	Archaic English 1.111111ft
Belgic (Doric) 1.07386ft –	Nippur 1.125ft
Sumerian 1.090909ft –	royal Egyptian 1.142857ft
Saxon* 1.1ft –	royal Egyptian 1.152ft
Archaic English 1.11363ft –	Russian 1.16666ft
Nippur 1.125ft –	Samian remen 1.178571ft
	(5/4 of .942857ft)
Royal Egyptian 1.14545ft –	Roman remen 1.2ft (5/4 of .96)

It is informative that in both the √2 and the π relationships that the remen (or *palimpes*) replaces the true cubit in the chord as the radius module diminishes, and conversely increases from the foot to the remen in the arc.

The whole number solutions are often maintained in the above results through the manipulation of the modules by the fractions 441 and 176 – meaning they are not all Root to Root solutions; the majority of results as shown have just one side of the equation as "Root."

*The Saxon foot is marked with an asterisk and indicates that it is the only solution that does not use 22/7 as the pi ratio, but uses 864/275 to produce the correct module in the result. (This practice will be dealt with below concerning both the geographic measurements and canonical man). The fact that emerges from the above comparisons, that of both the Sumerian and Saxon yielding a royal Egyptian solution, indicates that the Saxon foot is truly a modified Sumerian measurement.

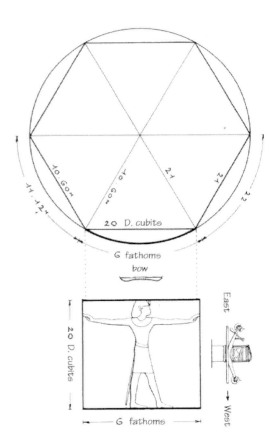

These same radii and the 60° arc relationships also hold good on greater modules, Schwaller noted the same connections as in the above diagram .

He gives the correct ratio of 21 radius to 22 of the 60° arc only in the bottom right section of the hexagon. If the six fathoms of the arc were six English feet, he may have noticed that the radius is then exactly the length of the King's Chamber in the Great Pyramid (34.3636ft is exactly 20 royal cubits). He uses "Denderah" cubits, a slightly longer version because his fathom is of one of the longer Greek feet.

The diagram below is of the feet and potential feet modules as shown in the original table above. It shows their relative lengths.

9	.9ft	Assyrian
10	.909090ft	Iberian
	.952714ft	unidentified
	.935704ft	unidentified (Chartres)?
15	.93745ft	Samian
20	.952381ft	lesser Roman
27	.964285ft	unidentified (Cycladic)?
48	.979591ft	common Egyptian
1	1ft	English/Greek
36	1.02875ft	common Greek
21	1.05ft	Persian
16	1.066666ft	Persepolitan
15	1.071428ft	lesser Belgic
12	1.090909ft	Sumerian
11	1.1ft	Saxon
10	1.111111ft	archaic English
9	1.125ft	Nippur
8	1.142857ft	royal Egyptian
7	1.166666ft	Russian/Jewish

The numbers to the left are the unit fractions by which they all vary from the English foot. Lesser modules than the English foot, such as the common Egyptian are 48 to 49 English, and modules greater than the English foot, such as the common Greek are 36 to 35 English – and so forth.

Two modules, the Samian and the Iberian, have been reduced from their true Root by a factor of 176/175 to produce this correspondence; this is the purpose of these fractions in the field of metrology – the maintenance of integers in designs or numerical series.

1.5 THE SHAPE OF THE WORLD

If not the sole reason, then at least one of the principle reasons for the regularly found variants of the modules, is the shape of the earth and how its oblateness affects the lengths of the meridian degrees. If the mass of the earth were spherical it would have the following geometric relationship with its satellite:

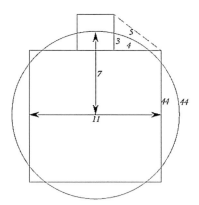

Squaring a circle to a ratio of 22/7 from the departure point of a 3-4-5 triangle

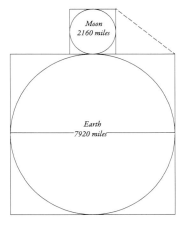

The Earth is not spherical and as such there are three radii that determine its oblateness – the polar, the mean, and the equatorial and their whole number relationships were taken to be as follows:

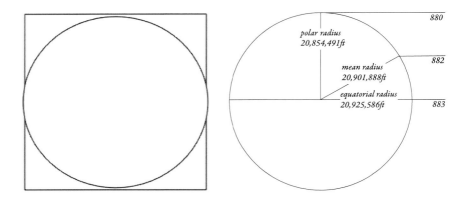

These lengths correspond to *one of the* modern acceptances of this magnitude:

EARTH'S RADII IN ENGLISH FEET

At the integer ratios:

880	882	883
Ancient metrology		
Polar: 20,854,491ft	mean: 20,901,888ft	equatorial: 20,925,586ft
WGS84 satellite		
Polar: 20,855,442ft	mean: 20,902,215ft	equatorial: 20,925,602ft
Correspondence %		
Polar: 99.9954%	mean: 99.9984%	equatorial: 99.99992%
Ancient Metrology		
6356.44km	6370.89km	6378.11km

The ancient estimates are as accurate as are the modern, this is because there is no current common agreement on an *exact* definition and diverse geoids, of which WGS84 is merely one, give different magnitudes within which these ancient estimates comfortably fall. The oblateness of the geoid that is due to its rotation, is not arbitrary in its numerical composition, it evidences a numeric-geometric regularity. This regularity could push the credibility dangerously close to creationism, but in conjunction with the specific patterns of metrology is better regarded as glimpses of the underlying symmetry of natural phenomena.

The meridian degrees of the Polar Regions and the degrees near the equator vary very little progressively from degree to degree – only about one metre per degree for three of the degrees at either end of the scale. From either end, in a regular fashion, this variation increases as they approach the median of 45° where the difference becomes *ca.* 20 metres between successive degrees. This allows the variable modules of antiquity to be assigned to very specific degrees on the surface of the spheroid. These diagrams of the quadrant arc of the earth explain certain of the variations that are universally found in the measurement system.

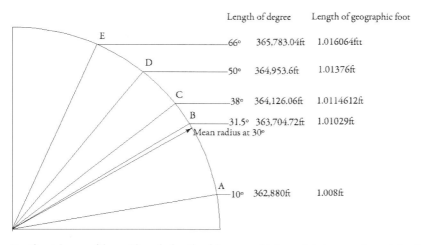

	Length of degree	Length of geographic foot
E		
66°	365,783.04ft	1.016064ftt
D		
50°	364,953.6ft	1.01376ft
C		
38°	364,126.06ft	1.0114612ft
B		
31.5°	363,704.72ft	1.01029ft
Mean radius at 30°		
A		
10°	362,880ft	1.008ft

Significant degrees of the meridian, the lengths of the geographic feet and their connecting unit fractions

A to B is 440 to 441	C to D is 440 to 441	D to E is 440 to 441
A to D is 175 to 176	B to E is 175 to 176	A to E is 125 to 126

What has been shown in the various tables are the "core" values that for reasons of neatness, if nothing else, merely make them more comprehensible. The additions to this core that are included in the above scale, is the northernmost degree of Egypt; this fits the pattern as being 441 to 440 of the least degree at 10° whose foot of 1.008ft is the first that achieves a geographic length, then at 1.01029ft is a localised Egyptian length and is not universally found. Also the northernmost degree at the latitude of the Arctic Circle where the foot length has increased to the maximum length and is twice the fraction of the least of 1.008ft, it is at 1.016064ft – 1.008.[2] This precise extended value is undoubtedly found in practise, but not as commonly as the core variants; it is the Greek foot of the Roman Pantheon design for example. (The full range of these variations will be demonstrated at a later point.)

1.6 Pi and the World

As was stated in the first diagram of this series (*see p.42*), if the world were a true sphere it would be 7920 statute miles in diameter. This is a radius of 20,908,800 feet, yet the mean radius of the oblate earth is 20,901,888 feet. The difference between these measurements is 3024 to 3025, the difference between the pi ratios of 864/275 or 3.14181 and 22/7 or 3.142857 is 3024 to 3025. In practice this means that the same meridian circumference *number* would be expressed if the pi value used on the perfect sphere were 864/275 and the universally used 22/7 were applied to the *mean* diameter, (here we show the radius): -

$$864/275 \times 20,908,800 = 131,383,296$$
(from the radius of the true sphere)
$$22/7 \times 20,901,888 = 131,383,296$$
(from the mean radius of the oblate at 30°)

The number 131,383,296 is the number of English feet in the meridian circumference that would be calculated from the mean radius. There are 129,600,000 geographic feet in this circumference; divide the former by the latter it equals 1.01376ft and this is the geographic foot calculated from the mean radius of the earth. Nowadays usually given as 309mm, it is one of the two most widely accepted values of a Greek foot, the other is 308.3mm and this is 1.0114612ft and both are sometimes termed "Olympic." The Parthenon was termed Hecatompedon which means hundred–footer, the stylobate width is 100 of these 309mm feet.

1.7 THE MERIDIAN CIRCUMFERENCE

Although there are many reckonings of the meridian circumference from the ancient world, the two most often referred to are in terms of the Greek 5,000 feet itinerary mile and the 6,000 feet nautical or surveyor's mile. The former begins as two remens - the step, two steps - the pace, one thousand paces - the itinerary mile; and the latter stems from two cubits - the yard, two yards - the fathom, one thousand fathoms - the geographic mile. The mile of the former is 5,068.8ft or 5,000 feet of 1.01376ft at 25,920 to the circumference; the mile of the latter is 6,082.56ft or 6,000 feet of 1.01376ft at 21,600 to the circumference.

SOME OTHER CLASSICAL ESTIMATES OF THE CIRCUMFERENCE:

- Eratosthenes 252,000 common Greek 500 feet stadia (36 to 35 of the "Olympic" Greek)
- Posidonius 240,000 Belgic 500ft feet stadia
- Ptolemy 24,000 Belgic 5,000 feet miles (66 ⅔ miles to the degree)
- Numerous sources give 27,000 Roman miles (75 Roman miles to the degree).

ARABIAN ESTIMATES:

- Habash 20,160 these are 6,000ft miles of the lesser Belgic feet
- Ibn Yunus 20,250 these are 6,000ft miles of the Persepolitan foot
- Other Arabian sources state the circumference is 66,000 farsakh (three miles to the farsakh or parasang) in which case this would be 19,800 miles of 6,000 Sumerian feet.

All of these estimates are accurate to the datum of 129,600,000 Greek geographic feet circumference.

1.8 THE HUMAN CANON AS THE SOURCE OF MEASUREMENT

The modules of measurement stem from the human form; wherever the skeleton articulates, it is a module of measurement. Modern antiquarians and archaeologists regard this human proportion as crude, "rough and ready" as it is often described. However, it would be better described prosaically as intelligent and convenient, or romantically as inspired and sublime. Undoubtedly, the best-known portrayal of the canon is Da Vinci's "Vitruvian Man" as shown overleaf.

Based upon the Roman architect Vitruvius' account of the principles of Greek architecture, in "The Ten Books on Architecture" Book III – "On Symmetry: in Temples and in the Human Body" he gives a description of the canonical scale between the various parts of the human anatomy. Although correct in every other detail, Da Vinci made one obvious correction to Vitruvius' description; this is the length of the anatomical foot in proportion to the stature. Given by Vitruvius as one sixth of the height, Da Vinci knew from his own draughtsmanship of the human anatomy that the correct proportion was one seventh and drew it accordingly.

Because canonical man is drawn in a square, then the circle of the same perimeter was superimposed upon Da Vinci's rendition; with the additional, previously described, means to draw the circle. Every aspect of the design proves to be a module of measurement from the datum of six Greek/ English feet as the height and reach of the man. The scale to the left is 144 half-inches and shows that the centre is taken to be the half-pi ratio 56 to 88 and not the phi ratio of 55 to 89. The perimeter of the square is 24 English feet and *Da Vinci's circle* is 24 Roman feet, the pi ratio is 864/275 in order to make the perimeter of the circle 24 × .96ft.

If the datum of six feet were *any* of the feet of the tables then comparable related modules would result. If similar data is applied to the scale of the woman one gets the same results from the underlying geometry. The basic rendition overleaf is by the Australian artist, Susan Dorothea White that she has entitled "Sex Change for Vitruvian Man" and the identical geometry has been superimposed upon it.

The rationale of the basic measurement is that in societies where nutrition is adequate and there is no sex discrimination, the ratio in stature between man and woman is 12 to 11. Therefore if the canonical man is six feet then canonical woman is 5½ feet. Through the application of the fractions 441 and 3025 as additional to the Root, then the woman is six Iberian feet in stature. It is gratifying to see the previously mentioned "cubits" of the ratios 3,4 and 5 English feet in practice. This how the units of measurements arise from the human anatomy.

The "Iberian" foot that arises from the female canon is no fudge of number, it is quite spontaneous and has (relatively) recent historical precedent. At .9166ft this is naturally 11 to 12 of the English foot and three such feet are 2.75ft. The Root Iberian foot is .9142875ft and if it is increased to the Standard classification by the additional 440th part it becomes .916364ft; it reaches the female canonical value of .9166ft by the addition of the 3024th part it then becomes rational to the English foot at eleven to twelve. (This is accomplished by using 864/275 as pi.)

IBERIAN FOOT (FT)	Root Reciprocal	Root	Root Canonical	Root Geographic
	0.909091	0.914285	0.919510	0.924765
	(27.710cm)	(27.867cm)	(28.026cm)	(28.187cm)

	Standard Reciprocal	Standard	Standard Canonical	Standard Geog.
	0.911157	0.916364	0.9216	0.926866
	(27.772cm)	(27.931cm)	(28.090cm)	(28.251cm)

Shown above are the core values of the Iberian foot that are subject to the same variations as the English/Greek. (It is the Standard value plus the 3024th part that is the basis of the female canon.)

In relatively modern times this identical adjustment was made to the Spanish vara (yard) after the annexation of the states of Florida, Texas, New Mexico and California by the United States. In Texas it was rounded to be 33⅓ inches but in New Mexico and California by 1854 it was rounded to be exactly 33 inches and this is 2.75ft, which divided by three, is the basic foot measure of canonical woman. Coincidentally, the measures used by the Spanish in Mexico were identical to the Aztec whom they supplanted, that were also based upon these feminine anthropometric values.

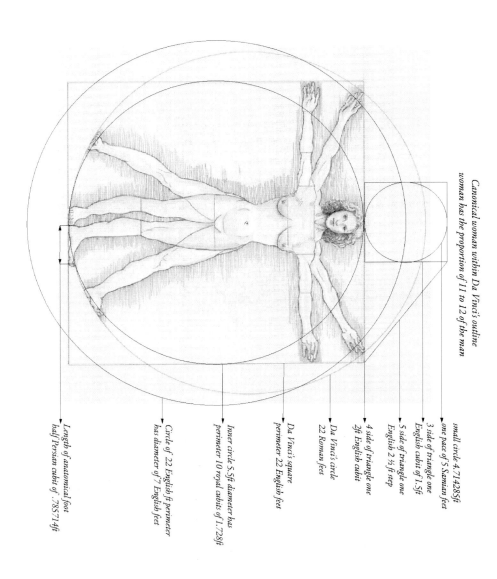

Canonical woman within Da Vinci's outline
woman has the proportion of 11 to 12 of the man

small circle 4.714285ft
one pace of 5 Samian feet

3 side of triangle one
English cubit of 1.5ft

5 side of triangle one
English 2 ½ ft step

4 side of triangle one
2ft English cubit

Da Vinci's circle
22 Roman feet

Da Vinci's square
perimeter 22 English feet

Inner circle 5.5ft diameter has
perimeter 10 royal cubits of 1.728ft

Circle of 22 English ft perimeter
has diameter of 7 English feet

Length of anatomical foot
half Persian cubit of .785714ft

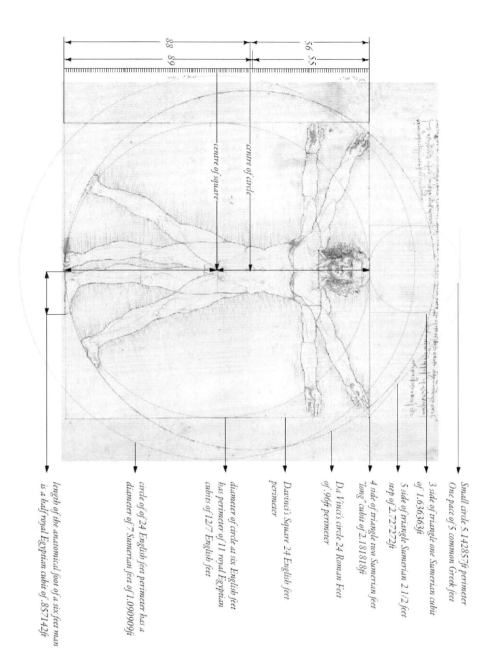

Small circle 5.142857ft perimeter
One pace of 5 common Greek feet

3 side of triangle one Sumerian cubit
of 1.636363ft

5 side of triangle Sumerian 2 1/2 feet
step of 2.727272ft

4 side of triangle two Sumerian feet
"long" cubit of 2.181818ft

Da Vinci's circle 24 Roman Feet
of .96ft perimeter

Da Vinci's Square 24 English feet
perimeter

diameter of circle at six English feet
has perimeter of 11 royal Egyptian
cubits of 12/7 English feet

circle of of 24 English feet perimeter has a
diameter of 7 Sumerian feet of 1.090909ft

length of the anatomical foot of a six feet man
is a half royal Egyptian cubit of .857142ft

Book II

MEGALITHIC MEASURMENT SYSTEM?

Preamble

It is decided to commence with this article because it was the first of its kind that I ever wrote, virtually at the point of the initial understanding of the unified system. This would be around 1996, it was originally entitled *Megalithic Measurement System* then it was emended around 2002.

In the light of subsequent findings and the growing disillusionment in the claims of Alexander Thom regarding measurement in general – the question mark was added to the title. The article is primarily the explanation of what the module "megalithic yard," if it existed at all, would be in metrological terms.

There should be modern technical emendations particularly regarding the Ring of Brodgar interpretation, the alternative and more probable solution to this stone circle is given in the later article "Measuring the Megaliths". Apart from the inclusion of illustrations it is presented as I left it at that time. This subject is now better understood and a more detailed assessment of the megalithic measures is given in the later article as mentioned.

MEGALITHIC MEASUREMENT SYSTEM?

2.1 THE MEGALITHIC YARD

It was in 1955 that Alexander Thom first proposed the existence of a rigidly maintained prehistoric system of measurements based upon a unit he called the Megalithic yard. He claimed that the Megalithic yard and associated modules governed the construction of Megalithic monuments throughout a wide geographic area. During 1968 for the first time in 100 years an ancient unit of measurement became internationally newsworthy, attracting both approving and condemnatory responses from the surprisingly wide spectrum of those who were interested.

This situation persists until the present day. For far too long, the existence or otherwise of this Megalithic yard has been an unresolved topic of debate among archaeologists. Despite the apparent dryness of the subject, the debate is often acrimonious. The orthodox view is that the idea of precision measurement among Neolithic and Bronze Age societies is preposterous; the progressive view is that a high level of sophistication governed both their science and social structure.

Because metrology has been more or less dropped from the studies of both archaeologists and the historians of science, there are no longer any experts to consult on the matter. This has led to the argument on the validity of the Megalithic yard being conducted by virtually unqualified people in this field. Representatives of either faction act out an absurd "Oh yes it is", "Oh no it isn't", circular impasse of academic bickering and the "evidence" called upon to support either view is scientifically flimsy.

The resolution of the validity of the Megalithic yard can only be reached via a proper understanding of the rules governing ancient metrology. Megalithic culture extends far beyond the regions explored by Thom and therefore overlap with ancient territories whose contemporary systems of metrology *are* understood.

2.2 The Metrological Background

Gaetano de Sanctis, a lecturer at the University of Rome during the 1930s, remarked: *"Ancient metrology is not a science, it is a nightmare"*.[1] The only leap of faith required of the reader is the acceptance that far from being a nightmare, the basic subject might be both simple and sublime. A plethora of modules were used concurrently in the ancient world and there are many differing values for each of them. At first sight the subject appears arbitrary, random and confused. The variable values have for long been regarded as evidence for slackness in the maintenance of standards, but this is not the case. Luca Petto, the renaissance antiquarian, remarked that if any of these variables were found to exactly agree, then this would imply an intended standard.[2] Seen from this angle, so many of the ancient modules are of precisely the same length that one must conclude that the variations are indeed deliberate. But what exactly are these variations – and what is their purpose?

W.M.F. Petrie was the most prolific, accurate and engaged researcher into ancient metrology. From his *Inductive Metrology*, (1877), to the publication of his *Measures and Weights*, (1934), he noted many of the variations evidenced by singular modules, certain of these variations were regularly of the 450th and the 170th part.[3] (As evidence now stands, these figures should be exactly refined as variations of the 440th and the 175th part). Petrie had reached his very accurate results by measurement alone, but the solution to ancient metrology is partially numerical in terms of certain absolute measurements, which leaves no tolerance at all in the definition and identification of related modules. That is, Petrie, and the majority of metrologists have had to express their given values between certain narrow parameters; but once certain measures are unequivocally identified – in an integrated system, they may *all* become expressible as an absolute value.

It is ultimately thanks to an insight of John Michell that it has been possible to make significant advances. In his *Ancient Metrology*, (1981), he perceptively established some of the governing principles that reveal the subject as a true branch of science. Much of his work concerns the geodetic appearance of certain of the modules, particularly the Greek, in that certain values of the Greek foot were sexagesimal fractions of the meridian degree. He was not of course the first to point this out, for the hypothesis has been regularly raised over the past 200 years and was widely stated to be so in classical antiquity.

Most importantly, Michell recognised that the ancient systems were most easily interpretable through the medium of the decimally expressed English foot. That is, fractional relationships in comparisons of differing measures are virtually undetectable when expressed in either millimetres *or* feet and inches, for this reason the systems of metrology have remained impenetrable. (Expressed in decimal English feet, one regularly sees the repetitive eleven fraction such as .1818 recurring or the septenary .142857, or 12 multiples such as 1.728 etc.). Most significantly he was also able to identify the fractional difference between singular measures as 175 to 176 — for example the Roman foot of .96768ft relates as 175 to 176 of the Roman foot of .9732096ft. Furthermore, he noted that this regular separation also occurred in the Greek, the royal Egyptian, the common Egyptian and sacred Jewish systems and modules. Thus, there were absolute values able to be expressed without margins of error, whereby one is able to detect direct linkages between cross-cultural systems.[4]

Livio Stecchini, probably the best-informed metrologist of the latter 20th century, noted many of these fractions linking ancient national standards. He graphically expressed certain of these links in the following manner:[5]

Mycenaean / Italic foot	15	9	
Roman foot	16		24
Greek foot		10	25

Had he continued to regard other systems from this point of view he might have recognised that they are all similarly connected. Furthermore, of the measures just listed, he never understood that it is the Greek foot that is pivotal to these interrelationships. He was thus unable to recognise the purely numerical structure of the metrological values that he could identify. From his wide experience and ability to make comparisons from an enormously broad array of values, he was able to very closely identify certain definitive values of the Roman and Greek feet. The Roman foot, he claimed, was exactly .9709501ft, and the Greek at the universally accepted 25 to 24 of this value he gave as 1.0114064ft. These values are *almost* exactly 440 to 441 of the values claimed by Michell, and for reasons adequately dealt with here it is the values proposed by Michell that proved to be precise.[6]

The higher value Michell stated of the Greek foot is 1.01376ft and 440 to 441 of this value is 1.0114612ft, there is little doubt, therefore, that the value of the Greek foot *almost* given by Stecchini is one of a series. Because in *numerical* terms of the English foot this shorter Greek foot is $(176/175)^2$, which is twice the fraction identified by Michell for differing values of the same module; obviously there are considerably more than the two values of each module he has identified. And, most importantly, what we call the English foot is shown to be one of the variables of the acknowledged series of the Greek foot: 1ft × 1.0057143 × 1.0057143 = 1.0114612ft.

It was the numerical structure underlying ancient metrology, had he but recognised it, that first attracted Algernon Berriman to the subject. (His *Historical Metrology*, (1953), for all its faults, errors and conjecture, remains the most frequently quoted authority on the subject of mensuration — (a fact which simply underscores the prevailing ignorance of this highly important branch of our historical inheritance). He noted that the sexagesimal 129.6 English inches as the perimeter of a circle, expressed a closely accepted value of the royal Egyptian cubit as the radius.[7]

Unfortunately he used true pi to divide the sexagesimal "canonical" number, 129.6, his resultant cubit thereby lost its precise definition. Had he used 22/7 as pi, a figure he clearly knew was an approximation used in ancient Egypt, then the cubit would have been the numerical absolute. The value arrived at by Berriman is 1.71887ft, but the precise value is 1.71818ft. This is exactly the length of the lesser value identified by Michell (related to a cubit as 175 to one of 176 equalling 1.728 ft). This is also the value arrived at by Petrie, principally from his examination of the king's Chamber of the Great Pyramid, but he was obliged to give the solution as 20.62 ± .005 inches; yet 1.71818ft is exactly 20.61818 inches.[8] (This point of canonical numbers as perimeters in terms of the English values will be laboured here because of its importance in the solutions of the stone circle designs).

If we apply the rule proposed in the explanation of the variations of the Greek foot and reduce the 1.71818ft cubit by its 441st part, then it is exactly 12/7 or 1.714285 English feet. This value has also been recognised as one of the standard expressions of the royal cubit. This royal Egyptian cubit is therefore exactly one English cubit of eighteen inches plus its seventh part. Consequently, since the English foot is one of a series of the Greek feet, then the seventh part added to any of the many values of the Greek cubit will equal a known value of the royal Egyptian cubit.

For many reasons, none of them partisan I hasten to add, it is the foot called English that is the basis or "Root" from which all calculations involving ancient metrology should begin. It would appear from most of the empirical evidence that the principal range of the variations in a single module, here given in terms of the variations of the Greek-English foot, are as follows:

GREEK/ ENGLISH	*Root Reciprocal*	*Root*	*Root Canonical*	*Root Geographic*
	0.994318	1	1.005714	1.011461
	(30.307cm)	(30.479cm)	(30.654cm)	(30.829cm)
	Standard Reciprocal	*Standard*	*Standard Canonical*	*Standard Geog.*
	0.996578	1.002272	1.008	1.01376
	(30.376cm)	(30.549cm)	(30.724cm)	(30.899cm)

For reasons dealt with elsewhere, the above terminology is used as descriptive in the classification of the variations. It was realised from the beginning that all of these variations were impossible to express in an ascending order. They must be tabulated in two rows, the fraction linking each of the variations across the rows is 175:176, and each of the values in the top row is linked to the value directly below as 440:441. "Root" prefixes the descriptive terminology from Least to Geographic in the top row and "Standard" in the bottom row. For example, 1.008 is *Standard Canonical* and 1.0114612 is *Root Geographic* etc.[9]

Because we are using the English foot to express these ancient measurements then as well as these values being measurements, they are also regarded as the *formulae* by which any other module is classified. For example, an established value of the Persian foot of Darius is 1.05ft. This is one and one twentieth of an English foot, therefore a *Root* value (at Root, the modules relate to the English foot by a single fraction, 1

1/10th, 1 1/14th, or 9/10 etc.). This Persian foot multiplied by 1.01376 = 1.064448ft which is the recorded value of the Arabian Hashimi foot and the basis of the French pied de roi, this would therefore be classified as a Standard Geographic Persian foot (the Persian being the "Root" value). No ancient modules are encountered that cannot be classified by the above methods

The principal reason that the regular interrelationship of ancient systems has not previously been recognised is that researchers have been trying to assess the elements at the wrong classifications. Thus they are viewing a compound fraction that *appears* to possess no mathematical relationship, (i.e. the fraction that separates the modules plus the fraction(s) of the module *variation*). The modules must be compared at the correct variation, that is, in the same column. For example, Freidrich Hultsch spent a lifetime attempting to link the common Egyptian foot of 300mm with the Roman foot of 296mm, but since the former is a Root Canonical (absolute 300.28mm) value and the latter a Root Geographic (absolute 295.96mm), it was an exercise in futility.[10] The ratio connecting the two *at the same classification* is exactly 49 Roman to 50 common Egyptian.

All of the ancient systems to which we erroneously give nationalistic names are similarly connected — by a unit fraction. This is in keeping with the ancient systems of mathematics that are a matter of record (multiple fractions such as 4/7^{ths} would have to be expressed as ½ + $^{1}/_{14}$th; and depending on the complexity of the original fraction, series of diminishing unit fractions approach the solution). This is how the units of ancient metrology fit together, the variables of the singular measures relate to their neighbours by a single fraction, and the measures relate to differing measures at the *correct classification* also by a single fraction. As the research progressed, such comparisons provoked the obvious conclusion — that the disparate "national" metrological structures were branches of a single system that was originally designed to be used concurrently.

The fractional difference of 175:176 serves the purpose of maintaining integers of the same module in both the diameters and perimeters of circles if the diameters are multiples of four. For example, in discussing the odometer, Vitruvius stated that a wheel of four feet in diameter travels 12 ½ feet in one revolution.[11] This is a very inaccurate pi ratio of 25/8 or 3.125, but is a value that is known to have been used throughout the ancient world. Far more commonly used is the quite accurate 22/7 or 3.1428571, and the difference between 25/8 and 22/7 is exactly 175 to 176. There is thus an eminently practical reason for this fractional separation in the modules: if Vitruvius' carriage wheel were four Roman feet of .96768ft (175) then the perimeter is indeed exactly 12 ½ feet, but of the related foot of .9732096ft (176).

The fraction 440:441 also serves the purpose of maintaining integers in diameters and perimeters *but of differing modules*, particularly if the perimeters are canonical numbers, i.e. sexagesimal and duodecimal solutions — numbers that are traditionally associated with circle perimeters. We saw this with Berriman's resolution of the royal Egyptian cubit, where the canonical number 129.6ins (correctly 10.8ft, also canonical) as a perimeter rightly yields a radius of 1.71818ft. Thus, there is a perimeter expressed in English feet and a radius expressed in royal Egyptian cubits, *but* the cubit related to the English is 12/7ft or 1.714285ft = 1½ ft + 1/7th, and this is the 440th part less than 1.71818ft.

The best example of this phenomenon is the division of the perimeter by 360; if this is regarded as feet, then the diameter is 114.5454ft, or 100 royal Egyptian feet each of (1ft + 1/7th) +1 /440th. Therefore the module of the perimeter must have the 440th part added to its related module of the diameter, so there is *Root* classification perimeter and *Standard* classification diameter, again, the separation of the 440th part is shown to be a corrective fraction related to the pi ratio and designed to maintain integers. This may be difficult to grasp but will be further explained as we analyse Megalithic circles. This rule is consistently observable with all of the canonical perimeter numbers in feet and the differing modules of the diameters.

2.3 THE VIGESIMAL COUNTING BASE

The last aspect of metrology to take into account before considering the Megalithic measurements is the nature of these modules other than feet and cubits. The basic unit of any system of measurement is the foot. It may be sub-divided into either inches, or with more versatility, into digits of which there would be sixteen, various multiples of these digits form other modules. Many of these modules persisted in the British imperial system into the 20th Century, but their antiquity was not remarked upon, being largely obscured by the sub-division of the foot exclusively into inches. On modern steel tapes the sixteen-inch division is still marked, but no longer has a terminology such as *pygon, remen* or *pygme*. The five feet *pace* was also commonly used until recent times. The *yard*, of course, is a *double cubit*, and a *cubit* is simply a device to alter the counting base to two instead of three, as in terms of the *yard* and *fathom*.

In Egyptian metrology there was a 20-digit measurement called a *remen*. The measurement systems of the Greeks and the Romans are directly related to the Egyptian and both had a 20-digit module, (this is not to say that the digit was the same length – they all differed by the same proportion as did their respective feet). In Greek, this unit was called a *pygon* and in Rome a *palimpes*. The *double remen* was therefore a *step (gradus)* of 40 digits and the double step was a *passus* or pace. As there are 80 digits in a pace, it is therefore five feet, each of sixteen digits. There are many digit multiples that were recognised as modules of measure. *Knuckle, palm, hand, lick, handlength, pygme, cubit* and so forth, and of these it is the handlength that interests us. From the heel of the hand to the fingertips this length, at ten digits, is a half remen, to this day, it is the basis of the counting system that is still used by the Welsh.

A twenty, or *vigesimal*, counting base is not peculiar to the Welsh, but was relatively common in the historical world; it is basically made up from two tens. In Welsh, twenty is *di-deg* – two tens; forty is then two two-tens, or *di di-deg*; sixty is *tri di-deg,* or three two-tens and eighty is *pedwar di-deg*, four two-tens, and so forth. That this method of counting – by the *score* – has survived from at least the Neolithic and Bronze Age to the present day should come as no surprise; our societies are permeated with ancient cultural distinctions that we unconsciously preserve.

As well as cultural differences in the ancient world, there were also very distinct cultural similarities and these are nowhere as obvious as in the counting systems and measurements that have survived from pre-history. If Alexander Thom had been correct he would have seen that the measurement system he proposed was used by the megalith builders, revealing them as "fully paid up members" of their contemporary world. This is because their method of measurement is identical in all respects to that of the majority of the systems of the ancient world; it is not unique to the Megalithic arena.

Alexander Thom, although he believed he had identified a module that had been consistently used in the construction of Megalithic monuments failed to *accurately* identify it. The modern thought processes require us to conform to a single standard; heretofore it has been the yard, rapidly being supplanted by the metre. One must not assume that the inhabitants of the ancient world viewed the subject of mensuration in this fashion. To them the subject was a science in its own right, rather than a simple method of quantification. It was Thom's modern way of thinking that forced him to try to identify a single module and an invariable value of that module. He mistakenly believed that basic measure to be the Megalithic yard, but this is essentially a compound measure (ie composed of more than one element). From his examinations of the monuments he identified rather more than the basic Megalithic yard, and taken altogether, these multiples and sub-divisions which he noted distinguish the branch of ancient metrology from whence they originate.

He noted that the stone circles, egg shapes and elliptical structures were often an odd number in terms of his Megalithic yard in their diameters; therefore there must be a half-yard module in the radius. A length of two Megalithic yards was detectable in the longer distances, which he termed a Megalithic fathom, and he claimed that the perimeters of the circles were designed to be in multiples of 2½ Megaithic yards, which he termed a Megalithic rod. If this were all of the information available, then the Megalithic series would have remained an irresolvable enigma, believable only by statistical mathematicians who would be unable to clearly demonstrate its existence to the layman, which has been the case. But Thom recognised another unit that seems to have escaped scrutiny and clearly identifies the system to which the measures belong.

The great stones are often patterned with "cup and ring" (*see overleaf*) markings which are thought to be contemporary with their host megaliths, and it was through analysis of their spacing and geometrical layout that Thom identified the basic sub-division of the longer modules. This was the Megalithic inch – forty to the Megalithic yard. With crystal clarity this enables the whole series to be categorised. All of Thom's appellations of his modules are misleading misnomers.

Cup and ring marks, Kilmicheal Glassery, Argyll

The "Megalithic inch" is clearly a digit, the half-yard is a remen, pygon or palimpes, the yard is a step or gradus, his fathom is the 5 feet pace and the rod is 10 palms, which suggests that the 10 digit palm would have been the basic module of the Megalithic engineers. Below is a table listing the Greco-Roman digit multiples:

MODULE NAME	DIGITS	FEET	MEGALITHIC EQUIVALENT
FINGER	1		Megalithic inch
Knuckle	2		
Palm	4		
Hand	5		
Lick	8		
HANDLENGTH	10		1/4 Megalithic yard (palmo)
Span	12		
FOOT	16	1	
Pygme	18	1.125	
PYGON (REMEN)	20	1.25	1/2 Megalithic yard
Cubit	24	1.5	
STEP	40	2.5	Megalithic yard
Xylon	72	4.5	
PACE	80	5	Megalithic fathom
Fathom	96	6	
(10 handlengths)	100	6.25	Megalithic rod
POLE	160	10	2 Megalithic fathoms

Note how the versatile digit structure allows for many counting bases, particularly duodecimal, but the Megalithic is clearly the remen or groups of twenty – two-tens, *di-deg*.

2.4 The Sumerian Cubit and the Megalithic Inch

When the Megalithic "inch" is compared with known ancient metrological systems at the commonly used multiples of 16 to the foot and 24 to the cubit, Thom's measurements are clearly what we call Sumerian measures.

Roman and Greek measures relate to each other by the ratio 24 to 25, Sumerian and royal Egyptian measures relate to each other by the same ratio, the Greek and royal Egyptian measures relate to each other by the ratio of 7 to 8. The Roman measures are therefore 7 to 8 of the Sumerian. In addition to the Megalithic yard, all these "systems" (among others) are significantly represented in British stone circles with the most minuscule margins of error; indeed, to ratios that far exceed the accuracy of 1:400 of which Thom claimed the builders were often capable.

Although no measuring devices survive from Sumeria, the varying values of the Sumerian cubit have been accurately established from the dimensions of buildings, the measurement of bricks, and from cuneiform tablets that record these dimensions. The most widely acknowledged value of the cubit is taken from that of the half-cubit represented as a rule on the statue of Gudea from Lagash, and given as 248mm. Thoreau-Dangin had previously given a related value of 495mm from cuneiform texts and plans of temple dimensions, which he subsequently excavated. As an absolute value it may be stated as 495.93mm, which is the classification *Root Least* (1.627066ft).[12] The other values of the Sumerian cubit drop into their proposed positions of the tabular arrangement – (at the ratios forwarded as above for the Greek feet) – with high degrees of accuracy, all the way to the maximum, *Standard Geographic* value, of 1.66836ft. This final measure is accurately given by the eastern side of the Ziggurat of Etemenanki at Babylon, as excavated under Robert Koldeway, and given as 91.52 metres, or 180 cubits, (calculated here as 91.532 metres).[13] These are the proposed values of the Sumerian cubit that are convincingly in agreement with the empirical evidence:

	Root Reciprocal	Root	Root Canonical	Root Geographic
SUMERIAN CUBIT	1.636363	1.645714	1.655118	1.664576
	(49.876cm)	(50.161cm)	(50.448cm)	(50.736cm)

	Standard Reciprocal	Standard	Standard Canonical	Standard Geog.
	1.640082	1.649454	1.65888	1.6683589
	(49.990cm)	(50.275cm)	(50.562cm)	(50.851cm)

Although the Sumerian cubit was most commonly sub-divided into 30 *shu-si,* Stecchini remarked:

> *"The texts do not draw any distinction between different types of cubits, except to state that the cubit usually divided sexagesimally into 30 fingers is at times divided into 24 fingers as in the rest of the ancient world".*

It is this 24th division of the Sumerian cubit that is the Megalithic inch. Although the royal Egyptian cubit had 28 divisions, they are not regular, the reasons for which have not yet been satisfactorily explained. However, Petrie, largely by his analysis of artist's grids that were accurately chalk-marked onto the walls of tombs, was able to state that they used a unit in multiples of 1/25th of a royal Egyptian cubit:

> *"Of these engraved lists the first two have a unit of a decimal division of the cubit; in No. 1 the spaces are 16/100 of a cubit wide, and 20/100 high, or 4/25 and 5/25; and in No.2 the spaces are 14/100 wide, and 16/100 high, or 7/50 and 8/50. The cubit of No. 1 would be 20.45 ± .05, and of No.2, 20.58 ± .08. This is of course inferior as a cubit standard to the determinations from large buildings but it is very valuable as showing the decimal division of this cubit, which is also found in other countries".*

> **"Pyramids and Temples of Giza"** *section 139*

(Additionally, Petrie has here accurately pinpointed two of the deduced values of the royal cubit, *Root Reciprocal* is 20.4545 inches and *Root* is 20.57142 inches).

Obviously this digit is the Sumerian, 24 to that cubit, 25 to the royal Egyptian and 40 to the Megalithic yard. More examples of the application of this Megalithic "inch" are cited below in the course of the Stonehenge description. But rather than labour through all the given values of the Sumerian cubit to substantiate the general theory, let us move directly on to the module which is the survival of the so called Megalithic yard into modern times.

2.5 THE SPANISH VARA

In his comparisons of the Megalithic yard with surviving measures, Thom gave much attention to the varied values of the Spanish vara.[14] After millennia of non-recognition, the most widely accepted value of the vara, that of Madrid and Castile proves to be the principal, or *Root* value, of the Megalithic yard. The exchange value of this vara is given as 2.7425ft and the value of a 40-digit module from the *Root* 24 digit Sumerian cubit is 2.742857ift, this is a ratio of accuracy in the region of 1:7,700. Even this negligible discrepancy is probably to be accounted for by the rounding down of a decimal.

The vara is divided into three feet of 12 *pulgadas* or inches, and in all probability these divisions were adopted at the time of the Roman occupation so as to come into line, so far as counting bases go, with the Roman *unica*. The alternative division of the vara is a length of 4 palmos, although these would each be of 9 pulgadas, the length would also be that of 10 Sumerian digits or Megalithic inches; ten digits being the palm length of the Greco-Roman measurement systems, and the basis of an essentially decimal digit count.

Having identified the Castilian vara as the Root value of the Megalithic yard, the related values may be expressed as follows:

CASTILIAN VARA (FT)	Root Reciprocal	Root	Root Canonical	Root Geographic
	2.7272727	2.742857	2.758531	2.774293
	(83.127cm)	(83.602cm)	(84.080cm)	(84.560cm)
	Standard Reciprocal	Standard	Standard Canonical	Standard Geog.
	2.733470	2.749090	2.768400	2.780598
	(83.316cm)	(83.792cm)	(84.271cm)	(84.753cm)

As well as regional variations of the vara within Spain, the Spaniards took them to all of the countries where they settled. Many of them subsequently fell from general use in the home country, but were preserved in the colonies. Thom noted some of these variations, and with additional variations from elsewhere they are tabulated below (derived from the Sumerian 40 digit module) as listed above:

Madrid	2.7425ft	correctly Root	2.74285
Burgos	2.766ft	correctly Standard Canonical	2.7648
Almeria	2.7329ft	correctly Standard Reciprocal	2.73347
Mexico	2.749ft	correctly Standard	2.74909
California	2.781ft	correctly Standard Geographic	2.78059
Texas-Peru	2.75ft	correctly Standard	2.74909
Canaries	2.7625	correctly Standard Canonical	2.7648

There are many more variations of the vara that accurately represent all of the proposed values and it is the value of the *Root Reciprocal* that was very closely identified by Thom from his surveys. He admitted that his definitive value of 2.72 feet was obtained by averaging, not realising that the variations that he witnessed were quite deliberate. The widespread practice of averaging in the study of ancient metrology has totally obscured its structure.

Land areas and itinerary distances also yield close values to those proposed in the above table. The length of the *legua,* a distance of 5000 varas, as used in the American Southwest is given as 2.6305 miles,[15] which differs from the *Root Canonical* by only 17.5 feet. The slight discrepancy may be accounted for by the acceptance of this value as 33 1/3 inches for convenience of conversion, instead of the 33.291 inches, which is the absolute. For a measure that is certainly of prehistoric derivation, it is a great tribute to the artisans, bureaucrats and scientists, who, over countless years, have maintained these standards.

2.6 THE SURVIVAL OF STANDARDS

Petrie speculated that constant copying had caused many of the variations within singular modules. Reasoning that standards over the years must lengthen, the copyists would err on the side of generosity because an error could be corrected were the rod cut too long, but would have to be discarded were it too short. This is obviously not the case. It was the custodians of the temples that manufactured and issued weights and measuring devices, as Petrie himself had established. Ritual stone rods were kept in the temples as standards from which others were copied or compared. Were these rods ever lost or destroyed, they could be reconstituted from the accurately known dimensions of the temple itself, which among its other functions, was regarded as the permanent repository of measure. Additionally, civic buildings would be of known dimensions and were often engraved with standards of measure upon officially approved stones as checks for market traders and artisans. By such methods, but principally by constant usage, standards remained unchanged for millennia. A new rod could be calculated from known constants – not copied from a worn out instrument.

These facts immediately answer the primary criticisms of Megalithic measures theory, notably by W R Knorr:

> *"How could such a unit be kept standard over more than a millennium and a geographic area of thousands of square miles? What would the standard be made of (wood? stone?) and where would it be kept? How could a population of dispersed migratory tribes maintain standardized measures, or even want to?"* [16]

The answer to the first question is that the units would be kept standard in the dimensions of permanent monuments. The second answer is stone, the stones of the monuments. The final criticism is specious, because to hypothesise that the Neolithic people were *"dispersed and migratory tribes"*, remains just that, a hypothesis.

The massiveness and uniformity of the megalithic structures that have survived indicate that these people were highly organised from centralised points throughout their territorial range. Prof. E.W. Mackie has astutely observed that the situation could not have come about unless there was a centralised training centre whereby instruction would be imparted in the necessary skills to achieve such uniformity.[17] He saw the whole scheme as analogous to the Roman Church, with its training colleges and rigidly hierarchical organisation. These centralised training points would be attended by the most able to receive instruction in the necessary skills of astronomy, geometry, geography and mensuration; of necessity, this would be accomplished under the umbrella of religion.

Although statisticians may be convinced of the regular unit construction of the Megalithic monuments, their analytical data is not comprehensible to the non-mathematician. This is because the analysts seem to show no particular necessity for the units that they identify within the monuments – and with which they were supposed to be designed – in any sensible integers within the constructions. This situation is entirely unsatisfactory; far better – John Michell:

> *"A tradition which has been credited by many learned men over the centuries is that the ancients encoded their knowledge of the world in the dimensions of their sacred monuments. If that is so, any attempt to elicit that knowledge must be preceded by a study of ancient metrology, for to interpret any set of dimensions it is of course necessary to establish the units of measure in which they were originally framed".*

Naturally, we should first establish the precise magnitude of that which is under investigation, and *then* seek the sensible integers that fit those measurements. Should those round numbers be discovered in modules that have been previously identified, these may then be considered to be a basis for a sensible theorem. Considered from this perspective, the Megalithic yard, fathom and rod at the values forwarded by Thom should, quite rightly, be dismissed. But if a consideration of the variables of the basic modules that he deliberated shows that they are present in the arrangements of the megaliths to great degrees of accuracy, this should help to convince the sceptical.

2.7 STONEHENGE

Stonehenge is the monument most amenable to such analysis. It is an intricate and unique structure, but to establish the regular modules used in its design it is not necessary to analyse every feature. The most important element of this complex structure is the sarsen lintel circle, which proves to be extremely informative. In its function as the repository of constant modules, a no more ingenious a device could be imagined.

The majority of the stones of Stonehenge are roughly shaped, but the imposts of the trilithons and the lintels of the sarsen circle are carefully dressed. The lintels are far above the ground and by this design have been preserved from damage and wear. They are carefully mortised and tenoned to the uprights and tongued and grooved together, this circle may therefore be calculated in both its intended inner and outer dimensions with the utmost accuracy.

Although the lintel circle itself is reduced to six stones of the original thirty, its inner diameter is identical to the inner diameter of the sarsen circle, the most accurate estimate of whose intended length is that of Petrie's examination of 1877.[18] He gives this dimension as 1167.9 ± .7 inches, which he identified as 100 Roman feet and since 100 Roman feet of the *Standard Geographic* classification (97.32096ft) is 1167.851 inches, it must be recognised as such. (That a skilled surveyor may deduce such information from a monument four millennia after its construction, should answer the question *"Where would such a standard be kept"*?).

Although the widths of the lintels have been roughly measured, no fully accurate estimate of the outer diameter of the lintels has been made until the publication of Michell's *"Ancient Metrology"*. He gave it as 104.2724571ft, an absolute that can be expressed within the parameters of the estimates of other surveys.[19] It is a correct solution for a number of reasons, not the least of which is the fact that temples were constructed to certain proportions whereby the modules of their design were deduced from *ratios*. It is significant that there are 30 lintels in the perimeter and that the width of the lintel is also one 30th of the overall diameter; the ratio of the inner to outer diameter is therefore 14 to 15. One fourteenth of 97.32096ft and one fifteenth of 104.2724571ft is 6.951497ft, and as a single module this is one Megalithic rod.

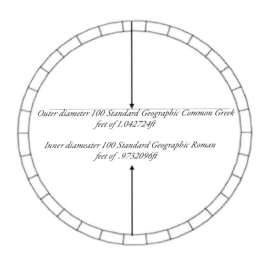

Outer diameter 100 Standard Geographic Common Greek
feet of 1.042724ft

Inner diameater 100 Standard Geographic Roman
feet of .9732096ft

The constituent Megalithic yard of this "rod" is the surviving length of the *Standard Geographic* Spanish vara whose absolute value is 2.780598ft. Because there are 14 rods in the inner diameter, there are exactly 44 in the inner perimeter, but the outer diameter being of 15 rods yields a perimeter that is not integral in terms of this module — there are 47.25 rods of the *Root Geographic* classification of 6.93573ft. Another solution must therefore be sought, in terms of a different module, for the outer perimeter.

The Megalithic yard in the guise of the Spanish vara also fits these diameters very informatively. Because there are exactly 35 in the inner diameter, it betrays the fact that there is a half yard module, (remen), in the radius. But most interestingly, there are exactly 37 ½ Megalithic yards in the outer diameter; therefore, as predicted from the appearance of the Megalithic yard module as a double remen, with a vigesimal counting base there would be a ten-digit palm in this radius.

The dimensions of Stonehenge incorporating Thom's Megalithic rod are entirely unsatisfactory. Although he had identified a value of the Spanish vara in excess of 2.78ft, it never seemed to occur to him to experiment with these increased lengths, as opposed to those that he thought definitive of the Megalithic yard module. He stated that the intention of the design was to contain the sarsens (or lintels) between concentric circles with circumferences of 45 and 48 Megalithic rods. Since the difference between the two is .48 of a rod of 6.806ft, this implies a lintel width of 3.26688ft, which as every survey of the stones reveals, is far too small.[20] He thus had to interpolate a whole rod into the inner circumference by using too small a value. The inner diameter he took as 97.41ft and employed a Megalithic yard of 2.7224ft, which fits this length in no sensible integer at all. It is small wonder that the rationalists have dismissed his results.

It is beyond dispute that Stonehenge in addition to whatever its other functions may have been, served as a repository of measures. The principal dimensions are designed in values of the *Standard Geographic* classification, (which means that in order to obtain the parity of the modules with the English foot, they must be divided by 1.01376, which reduces them to *Root*). Exactly how all the measurement systems of remote antiquity once formed an integrated system is beautifully exemplified in Stonehenge with some of the more interesting modules:

The inner diameter	100 Roman feet of	.9732096ft
	98 Common Egyptian feet of	.993071ft
	96 Greek feet of	1.01376ft
	56 Royal Egyptian cubits of	1.737874ft
	35 Megalithic yards (varas) of	2.780598ft
	14 Megalithic rods of	6.951497ft
The outer diameter	105 Common Egyptian feet of	.993071ft
	100 Common Greek feet of	1.0427245ft
	90 Royal Egyptian feet of	1.158583ft
	60 Royal Egyptian cubits	1.737874ft
	50 Jewish sacred cubits of	2.0854491ft
	37.5 Megalithic yards (varas) of	2.780598ft
	15 Megalithic rods of	6.951497ft

Interesting connections become clearer by such methods of tabulation. Frank Skinner, who had charge of the Weights and Measures Department of the Science Museum during the 1950s, noted that the common Greek foot related as 3/5[ths] of the royal Egyptian cubit, and since the sacred Jewish cubit relates to the royal Egyptian as 6 to 5 it is clear that the common Greek foot is a half sacred cubit.[21] More such loops can be easily spotted in the above table. The sacred Jewish cubit (that of Moses and Ezekiel) expressed as "a cubit and a hands breadth" refers to the royal Egyptian as the basic cubit. Since a hands breadth is 5 digits, the implication is that 5 such hands comprise the royal Egyptian and 6 the sacred Jewish, this digit is therefore the Megalithic "inch": 15 to the "common" Greek foot, 24 to the Sumerian cubit, 25 to the royal Egyptian cubit, 30 to the sacred Jewish cubit and 40 to the Megalithic yard. There is little doubt that we are regarding a single organization of measurement in what has previously been viewed as quite separate systems.

One of the most intriguing solutions to these numerical harmonies is that of the outer circumference of Stonehenge, 327.713ft, being exactly 360 Italic or Mycenaean feet of .910315ft as an absolute. This canonical perimeter number is of the *Root Geographic* classification, the 440th part less than the *Standard Geographic* measures of the diameter, this is the corrective fraction, previously mentioned, that maintains integers and happens arithmetically. There are 100 common Greek feet in the diameter and the common Greek foot is a Mycenaean foot plus the seventh part.

The modules of diameters are therefore composite measures if the perimeters are basic foot measures in canonical multiples. A pertinent example is to view the compound module of 100 *Standard* Megalithic yards as a diameter of 274.90909ft, the perimeter is then the canonical number 864 English feet. It is to be noted that the composite measures of the diameters are invariably expressed decimally and the perimeters duodecimally. These dual counting bases in harness have other metrological properties, but would be digressional to explore at this point. It would thus appear that the system had been observed or *discovered* through simple arithmetic, rather than contrived.

Stonehenge is often described as a great temple with a surrounding necropolis of the burial mounds of Neolithic Bronze Age royalty or heroes. The monument is supposedly unique, with no ancestors or descendents. But a metrological examination of other Megalithic monuments reveal similar solutions in precisely defined values to such degrees of accuracy that they may only be recognised as intentional, and thereby directly related to the design of Stonehenge.

Without considering the complexities of the repetitive shapes of the rings identified by Thom, the existence of a regular measurement system is more simply demonstrable if confined to the straightforward analysis of true circles.

2.8 ROLLRIGHT, OXFORDSHIRE

The Rollright Stones

The Rollright Stones of Oxfordshire[22] have a very obvious and direct metrological relationship with Stonehenge. Although many of the stones had been removed early in the 19th century they were replaced in 1866 when such "restorations" were fashionable, and their intended dimensions may still be accurately assessed. Given by Thom as 103.6ft, the diameter is within .96 of inch of its presumed value of 103.68ft; because 103.68ft is exactly 175 to 176 of the outer diameter of Stonehenge at 104.27245ft. The modules in the Rollrights are therefore all calculated at the *Standard Canonical* classification of measures, 1.008 greater than *Root*. All the elements which fit the Stonehenge dimension therefore fit the Rollrights *at this classification*. This indicates a margin of error in the region of one part in 1300, which of course is no error at all; any surveyor would allow such latitude. The Megalithic yard of this classification is that of the vara of Burgos, 2.7648ft, at 37.5 in the diameter; and if one uses that measurement here it is likely to be in counts of the 10 digit palmo at 75 in the radius, or, as at Stonehenge, 15 Megalithic rods diameter. However, the most likely intention at the Rollrights is 100 common Greek feet diameter and 360 Mycenaean feet perimeter. Thom claims that the diameter is 38.1 Megalithic yards, this yields 2.71916ft, which is numerically unsatisfactory and unlikely as part of any design.

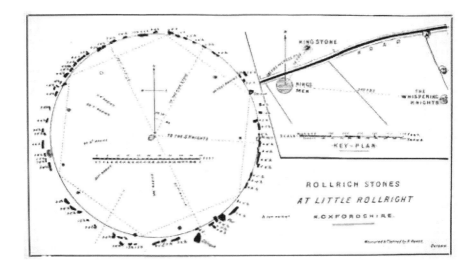

2.9 THE MERRY MAIDENS, ST BURYAN, CORNWALL

The Merry Maidens at St. Buryan is another true circle, also connected to the Stonehenge measurements in as much as it is constructed with the maximum, *Standard Geographic* modules. Thom remarks that it was re-erected, but it comprises stones of such sturdy simplicity that it is certain that it reflects the original design. He gives the diameter as 77.8ft or 28.6 Megalithic yards and perimeter as 35.9 Megalithic rods, which is an arbitrary Megalithic yard of 2.72027ft, and it is little wonder that his reasoning of a constant unit is disputed.[23]

The diameter of this circle is 80 Roman feet of .9732096ft to within less than ¾ of an inch, but because this is a multiple of four it is unlikely that this is the module with which to interpret the metrological design. (The four multiple cannot give an integral perimeter by the established method of adding the 175th part, because at the *Standard Geographic* this classification is already the apparent maximum). 80 Roman feet is, however, also 70 Sumerian feet, which yields an eminently suitable 220 Sumerian feet of *Standard Geographic* of 1.11224ft as a perimeter. This is also 88 Megalithic yards of 2.78059ft (the vara of California) and 35.2 Megalithic rods.

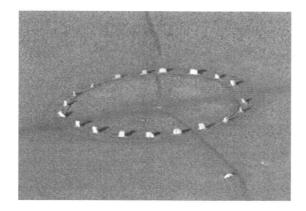

The Merry Maidens (pictured above and below)

Both 88 and 352 are significant in canonical number theory, and it would seem from this observation that Thom was correct in much of his reasoning; in this case, in his claims that the designers were obsessed with maintaining perimeter integrity.

Thom believed that they "sacrificed" diameter integrity to sustain this; but with a variable base measure both may be obtained with exactitude. Again, it would seem from this interpretation of the Merry Maidens that the addition of the seventh part to a module has a function related to *pi* in the maintenance of integers. (Mycenaean plus the seventh is common Greek, Roman plus the seventh is Sumerian and "geographic" Greek - of which the English is a component - plus the seventh is royal Egyptian).

Perhaps the most widely acknowledged connection between what are often wrongly believed to be separate systems of measure, is that of the Roman to Greek at 24 to 25. A Roman furlong of 625 feet is the equivalent of a Greek stade of 600 feet. This too, is related to the pi ratio. One Roman foot radius is six Greek feet perimeter. But once again, there is a classification change in the modules; the foot of the perimeter is the 175ᵗʰ part greater than its relative of the radius. (Berriman noted this exact relationship). Therefore, one Sumerian foot as a radius is six Egyptian feet as a perimeter. Six royal Egyptian feet (or four royal cubits) is therefore 6.25 Sumerian feet, which is exactly the Megalithic rod (2 ½ Megalithic yards which are each 2 ½ Sumerian feet).

2.10 The Ring of Brodgar

The Ring of Brodgar on mainland Orkney provides a perfect example of the kind of ambiguities thrown up in the interpretation of measurements. Thom found the diameter to be 340.66 ± .44ft and rightly gave the distance as 125 Megalithic yards. However, he took the lesser estimate and took the Megalithic rod to be 6.813ft, stating:

> *"The reason for using a circle diameter 125 Megalithic yards is that it gives a perimeter of 392.7 Megalithic yards which is very nearly an integral number of Megalithic rods".* [24]

Of course, it is nothing of the sort; at 157.08 Megalithic rods it is quite meaningless. The correct interpretation of the geometry of Brodgar, however, yields perfect integers in previously identified modules. The diameter is intended 125 Megalithic yards of the *Root Reciprocal* 2.7272ft and 340.909ft is perfectly within the measured length. As well as being exactly 50 Megalithic rods it is also 200 royal Egyptian cubits. Because both are a decimal count the resultant perimeter will be in modules of the *Root* classification, the 175ᵗʰ part longer than those of the diameter. The ensuing perimeter is 1071.4285ft and the temptation is to leave it there, in that it is 1000 feet of 1 1/14ᵗʰ English feet, and although this module is difficult to categorise, similar and related lengths are reported from many ancient sources. It is in perfect agreement with other aspects of metrology by being at *Root* an increase of the English foot by a single fraction. At certain of its values it overlaps, exactly, the so-called Germanic or Drusian foot, which is 18 digits to the Roman 16.[25] In Greek terminology this Germanic foot would be termed a *pygme* at its Roman reduction.

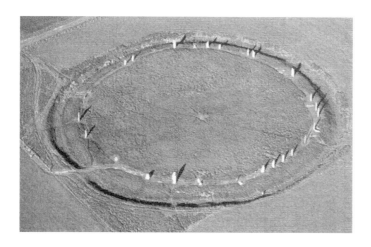

The Ring of Brodgar, Orkney

The perimeter of Brodgar is also exactly 625 Royal Egyptian cubits of 12/7 English feet, but this is a numerically unsatisfactory number for a circumference. Certainly, this length cannot be interpreted in terms of either the Megalithic yard or the Megalithic rod. *Note. (Thom claimed that it has exactly the same diameter as the two inner circles of Avebury, but this is not the case).*

This gives credence to the claims of cultural regularity and wide dispersal of monuments that had an obvious scientific purpose, if only as an exercise in geometry. Obviously it was much more. This is apparent by what we know of the enormous difficulties that were experienced in the development of the metric system, which can only be used to quantify. In comparison to the system of antiquity, it is childishly simplistic.

(Emendation note: In the course of time these Brodgar dimensions have been refined. Dealt with in the article "Measuring the Megaliths.")

From these few examples we can see how modules that have been precisely defined elsewhere, fit British Megalithic geometry without margins of error. The plausibility of this argument is supported by the fact that the modules that have been identified also fit the monuments in rational sets of numbers.

Such is the integration of metrology. Are we then dealing with a "Megalithic" system of measures? Probably not – nor any other that could be labelled. It would seem that all of the systems which we brand with a nationalistic nomenclature are simply contrived from singular standards which have been drawn from an older complete cosmology by the bureaucrats who sought to quantify, tax and regulate the societies into which it was fragmented. Were it not for the survival of the vara into modern times then Thom's claims of a consistent unit would have remained forever enigmatic. For all of the evidence that Thom brings to bear on the subject, from his own statistical methods to those of Broadbent, Kendall and Freeman, whom he cites, they remain virtually useless in the definite establishment of the Megalithic yard. Because for all the fine terminologies in which statistical methodologies are cloaked, they simply remain other methods of averaging. Averages are the enemy of precise definition and in most cases are misleading; they have certainly obscured the clarity of ancient metrology.

Although the few examples that have been included here are persuasively indicative of a regular and systematic usage of a single system across a wide geographic area, only around 30% or so of Megalithic monuments (apparently) conform to the "megalithic" numerical scheme as outlined. So the argument for the Megalithic barely yard remains open, but a greater beast has been unleashed.

SOURCES

[1] Stecchini, L.C. **http://www.metrum.org/measures/whystud.htm**
[2] **http://www.metrum.org/measures/romegfoot.htm**
(Few works are published by Stecchini, but his memorial web site is the equivalent of several books)

[3] Petrie, W.M.F. *Encyclopaedia Britannica*, Eleventh Edition, 1910, (Ancient Historical) Weights and Measures, p 481

[4] Michell, J. 1981 *Ancient Metrology*, p 17

[5] Stecchini, appendix to Tompkins, P. *Secrets of the Great Pyramid*, 1971, p 352

[6] Neal, J. *All Done With Mirrors*, 2000, pp 69-75

[7] Berriman, A.E. 1953 *Historical Metrology*, pp18-19

[8] Petrie, W.M.F. 1883 *The Pyramids and Temples of Gizeh*, ch 20, para 137

[9] Neal, J. *All Done With Mirrors*, 2000

[10] Stecchini, appendix to Tompkins, P. *Secrets of the Great Pyramid*, 1971, p 309

[11] Vitruvius, *On Architecture*, Book 10 ch. 9 (Trans Frank Granger, 1934, from the Harleian manuscript).

[12] Skinner, F. *Weights and Measures*, 1967, HMSO Science Museum, p 41

[13] Stecchini, L. **http://www.metrum.org/measures/length_u.htm** para 7

[14] Thom, A. & Thom, A.S. 1978, *Megalithic Remains in Britain and Brittany*, p 43

[15] Rowlett, R. How Many? *A Dictionary of Units of Measurement*, (revised 2001) University of N. Carolina at Chapel Hill

[16] Knorr, W. R. *The geometer and the archaeoastronomers: on the prehistoric origins of mathematics*. Review of: *Geometry and algebra in ancient civilizations* [Springer, Berlin, 1983; MR: 85b:01001] by B. L. van der Waerden. British J. Hist. Sci. **18** (1985), no. 59, part 2, 197--212. SC: 01A10, MR: 87k:01003.

[17] MacKie, E. 1977, *The Megalith Builders*, Phaidon Press

[18] Petrie, W.M.F. 1877 *Stonehenge: Plans, Description, and Theories*. p 23

[19] Michell, J. 1981 *Ancient Metrology*, p 20

[20] Stone, E. Herbert. 1924 *The Stones of Stonehenge*, p 6

[21] Skinner, F. *Weights and Measures*, 1967, HMSO Science Museum, p 35

[22, 23, 24] Thom, A. and Burl, A. 1980, *Megalithic Rings*

[25] Skinner, F. *Weights and Measures*, 1967, HMSO Science Museum, p 40

Book III

THE METROLOGY
OF THE BROCHS

THE METROLOGY OF THE BROCHS

The following was the first specialised article to be written post publication of *All Done With Mirrors*. The writing was inspired by the fact that the subject matter concerned a cluster of objects of a similar nature that would most likely have incorporated regular measurement within their design, in this case they are, predominantly Iron Age, circular and fortified stone buildings. Dr Euan Wallace MacKie supplied the data on these buildings.

Dr MacKie is highly esteemed by progressive researchers, he has written extensively on archaeoastronomy and the megalithic culture and has thoughtful, well-considered opinions on these subjects. He began publishing his work in 1961 and continues to the present. He sent to me the dimensions of 49 of the Scottish brochs and by applying the simplest of techniques on the data I was able to identify certain regularities within their design. This method of module analysis was simply to divide the diameters by a multiple of seven, an approach that had consistently yielded logical results during previous research on megalithic circles.

Although the article is largely in its original form certain emendations have been made to modify it to the standard of knowledge subsequently discovered and presently extant. In the original, certain of the broch dimensions, namely those of Dunrobin Wood, Brae and Dun Boreraig, had been presumed to have a multiple of four as a diameter because the division by seven had not yielded a recognisable module. Such emendations are inserted into the original text with appropriate notes.

This division of a diameter by four had been practised since antiquity in order to produce integer solutions in both diameter and perimeter; the pi ratio used in this practise is 3.125 or 25/8. The difference between this ratio and the commonly used 22/7 is the unit fraction 176/175. Because all modules in the ancient world have regular variations in their values of this precise fraction, then the module of the perimeter being this exact amount longer than the module of the diameter maintains the accuracy of 22/7. The complex system of metrology appears to be designed around the necessity to maintain integers.

In the intervening years other modules have come to light that fit the spaces left in the majority of the brochs to make all of them conform to multiple seven diameters. After much experimentation, which resulted from John Michell's astute observation that many modules of foot lengths related, not only by unit fractions such as 24 Roman to 25 Greek, but that these cross-cultural fractions were largely "square numbers," 25 being the square of five, in this case. I had identified many different national foot values that John had not considered in his researches, and pieced them together in a tabular form that illustrated their unit fraction connectivity. For example Assyrian measure is 9 to 10 of the Greek, Roman is 24 to 25 of the Greek and the Assyrian is then 15 to 16 of the Roman. Add the Belgic, which is 9 to 8 of the Roman and it is also 6 to 5 of the Assyrian, and so on and so forth.

John took my tabular arrangement of 12 distinct feet of different nations and arranged them in an ascending order through "square numbers"; in order to make this cohesive table he had to propose not twelve, but nineteen *potential* foot values with the term "unidentified" appended to at least six of them at that time. He started with least foot value, which was the Assyrian .9ft and noted that it was 63 to 64 of its neighbour the Iberian foot of .9142857ft, (64 being the square of 8). The relationships continue through 81 to 100 to 121 then descend again including 49, 36, 25, 16, and 9, ending at the maximum value for a foot which is the Russian foot at 1.166666ft or 7/6 English feet. In the fullness of time very positively identified foot lengths that we had not previously recognised filled the "unidentified" spaces in the tabular form. This will be illustrated several times in the ensuing articles.

One value in particular that exemplifies this coming to light is the foot termed Samian; identified as a number that was related as 99 to 100 of the lesser Roman foot. Its value is .9428571ft, this (and its variants) was found to be the city standard of no less than 12 pre metric German cities and was also encountered in Gothic cathedral construction. It was first detected by the fact that according to Herodotus there are 800 feet in the Great Pyramid base, this would be a foot of .945ft. All of the modules in the GP are the 440th part greater than the Root values of the tabular arrangement. Therefore Herodotus' foot is that of .9428571ft at its "Root"; it was called Samian as Samos was the birthplace of Herodotus. It is this foot that was found to correct the broch diameters of Dunrobin Wood, Brae and Dun Boreraig to being seven multiples in diameter, thereby unifying all of them into the same formula.

In all probability, the measurements as supplied by Doctor (now Professor) MacKie may not be as accurate as inferred in this article, but the only error can be in the exact classification; it is merely an excercise in the application of metrology regarding technique and module.

3.1 THE BROCHS — 49 MEASURED SITES

Throughout Scotland and the Scottish islands there are in excess of 200 and possibly 500 major broch sites. The following analysis is taken from, what I believe to be, the accurately measured inner diameters of 49 of them as supplied by Dr Euan MacKie. The modules are expressed in English feet although the original measurements were taken in metres, for reasons of ease of analysis they are converted to feet at the rate of 3.2808427 feet to the metre. The range of diameters extends from the smallest, at Mousa, 18.897654ft, to the greatest at Oxtrow at 44.816311ft.

A Hebridean broch showing
likely roofing arrangement.
Massive walls were
hollowed into chambers.
Reconstruction by
Alan Braby

The evidence would imply that a professional class of masons were employed in their construction throughout the extent of their range and the time span of their unique design. The system of measurement employed in the brochs, both in the module lengths and the methods of application, is identical to that of the preceding megalithic – Neolithic/Bronze Age societies; *and* to the cultures that succeeded them. The most interesting fact that clearly emerges from the cumulative evidence is that the builders applied certain formulaic procedures in their plans. The vast majority of the diameters are multiples of seven in terms of the various feet that are used; and these diameters *become* exactly seven when known module multiples of these feet are employed. What is meant by this, is that diameters that are 21 feet would be seven yards (ancient metrologists sometimes expressed the yard as a double 1½ ft cubit); if they are 28 feet they are seven double two-feet cubits (bracciae); at 35 feet they are seven five-feet paces (double step) and those of 42 feet are seven fathoms

As the size of the brochs increase, the *numbers* of the modules do not; the modules themselves increase in order to maintain the numerical formulae. The first six brochs of the list illustrate this point at their *theoretical* values with their variance from the *measured* length being quite minute:

Mousa	18.9ft =	21 Assyrian feet of .9ft.
Nybster	21ft =	21 English feet.
Ousedale	21.84756ft =	21 common Greek feet of 1.04036ft
Castle Cole	22.176ft =	21 Persian feet of 1.056ft.
Armadale	22.94ft =	21 Belgic feet of 1.092378ft
Dun Carloway	24ft =	21 royal Egyptian feet of 1 14285ft

The following 20 brochs, with a couple of notable exceptions, from number 7 in the list, Kiess North, at 28.8934ft to number 27, Clachtoll, at 31.36ft, have diameters that are each of 28 feet which range from Iberian to archaic English.

From number 28, Midhowe, to number 31, Loch of Huxter, each are 35 feet of the Assyrian variants; the diameters of the next three, from 32 to 34, revert to being 28ft of the greater measures, royal Egyptian and Russian. From numbers 36 to 46 the diameters are again of 35ft in terms of the range of possible measures between the lesser Roman values ascending to the greater values of the royal Egyptian. Finally, when the diameters exceed 40 whole English feet, the division of the final three brochs of the list, are in terms of 42 feet of the common Egyptian and the Persian standards

The dimensions of the brochs with the measured values and the theoretical absolutes are as follows:

Site name	D. feet	P. feet
1 Mousa	18.897654	59.392627

This diameter has a clear resolution in terms of the Assyrian Root foot of .9ft at 21 or seven yards.

Mousa, one of the better preserved brochs

2 Nybster	20.997393	65.991807

Obviously seven three-feet yards diameter, perimeter 22 yards or 66ft. Module – English foot.

3   Ousedale Burn   21.850412   68.672725

The module here is the Common Greek "yard", each three feet of 1.04036ft (Root Geographic) and 22 such yards perimeter or 66 common Greek feet.

4   Castle Cole   22.178497   69.703847

Seven yards of The Persian foot of 1.056ft, (the 5000th part of the Statute mile). Again, 22 such yards or 66ft perimeter. This foot is often encountered in the Gallic 7500 feet leagues.

5   Armadale Burn   22.965899   72.178539

Obviously this diameter is seven "metres". The metre is composed of three Belgic feet; it is Root Geographic at 3.277134ft the constituent foot being 1.092378ft.

6   Dun Carloway   24.015769   75.47813

One seventh of this diameter (24ft) is three royal Egyptian feet of 1.142857ft.

7   Keiss North   25.918657   81.458637

25.8934ft is 28 feet of the Root Geographic Iberian foot, this is within 1/3 of an inch of the measured length.

8   Dunrobin Wood   26.377975   82.902208

This is eight yards composed of what are termed Sumerian feet of 1.099636ft, yielding a perimeter of 75 Sumerian feet of 1.10592ft, 50 cubits of 1.65888ft or 30 steps of 2.7648ft. This is also the Spanish vara of Burgos, therefore it additionally has a three feet subdivision, so this perimeter could also be viewed as 90 Iberian feet.

9   Brae   26.574826   83.520881

The same solution must be forwarded here as that of Dunrobin Wood, whereas the margin of error at that site is around one part in 2000, the margin here would be one part in 800.

10   West Burra Firth   26.90291   84.552003

11   **Borwick**   26.90291   84.552003

The most likely solution here is a Roman foot module of .96ft because these feet at 28 to the diameter equal a rational 88 feet perimeter, which is a module number often encountered in the older megalithic monuments. The margin of error is ¼ inch overall.

12   **Backies**   27.099761   85.170676

This too would have the 28 Roman feet solution, very accurately at 99.983 percent in terms of the Standard Canonical Roman foot of .96768ft.

13   **Dunbeath**   27.821546   87.439145

With less than a quarter inch error on the diameter, this is 28 Standard Geographic common Egyptian feet of .993071ft (6 sevenths of the royal Egyptian foot). Therefore giving the same numerical solutions as the preceding five circles. Many other modules are compatible with this diameter, it is ten Spanish varas as used in California, 25 Sumerian feet and 24 royal Egyptian feet.

14   **Levenwick**   27.95278   87.851594

It is difficult to see this as anything other than an intended 28 English feet; it is about ½ inch short.

15   **Dun Troddan**   28.084014   88.264042

As above applies to this circle, but in this case it is an inch too long.

16   **Brounaban**   28.14963   88.470267

This measure is identifiable as a Root Canonical Greek foot of 1.0057142, as identified by Martin Folkes in 1736; he had noted it as being engraved on a Standards Stone at the Roman Capitol. At 28 to the diameter it offers the same *numerical* interpretation to the previous circles, as do the following.

17   **Dun a' Chaolais**   28.8058   90.532511

At exactly 28.8 feet this is 28 common Greek feet at Root Classification.

18   **Jarlshof**   29.461967   92.594755

29.4668ft is 28 Standard Persian feet of 1.052386ft.

19 Howe of Hoxa   29.724435   93.419652

This length is about a tenth of an inch short of 28 Root Geographic Persian feet of 1.062034ft.

20 Kylesku   29.986902   94.24455

In this case, the diameter is 28 Root Belgic feet of 1.071428571feet, again with about a tenth of an inch in excess in the measured overall distance.

21 Clickhimin   30.24937   95.069448

28 Standard Canonical Belgic of 1.08 feet would be 30.24ft.

22 Burrian   30.77430   96.719243
23 Carrol   30.77430   96.719243
24 Caisteal Grugaig   30.77430   96.719243

These three are within a fifth of an inch overall of 28 Standard Sumerian feet (the basis of the Saxon or Northern foot). Additionally, the perimeters would also be 100 Standard Canonical Roman feet of .96768ft.

25 Kintradwell   30.97115   97.337916

This is accurately 28 feet of the Standard Canonical Sumerian foot of 1.10592ft, this perimeter taken as 97.32096ft is the precise inner diameter of the Stonehenge Sarsen circle; 100 Standard Geographic Roman feet.

26 Dun Fiadhairt   31.233623   98.162814

This distance at a little over half an inch either way would make it either the Standard or Root Canonical classification of the archaic foot of the "yard and full hand" at 1.11111ft.

27 Clachtoll   31.364856   98.575262

This is very clearly 28 of the Standard Canonical classification of the same foot, at 1.12ft.

28 Midhowe   31.627324   99.40016
29 Borrowstone   31.627324   99.40016

Were these circles about half inch longer at 31.68ft, when divided by the next multiple of 7 at 35, it equals the Root Canonical value of the Assyrian foot of .905142ft; thereby making the perimeter rational to the same classification of this foot at 110. This number lends itself ideally to the 2½ ft "step" division at 44 in the perimeter.

30 **Yarrows**   31.889791   100.22506

This circle offers the same numerical solutions in terms of the Root Geographic classification of the Assyrian foot, 35 of which equal 31.861ft.

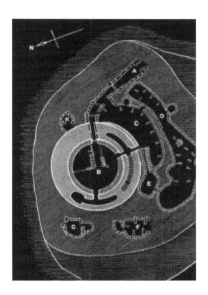

Plan of Yarrow Caithness, the north east coast where there are many brochs

31 **Loch of Huxter**   31.955408   100.43128

The Standard Geographic value used in the same way would yield a diameter of 31.9334ft.

32 **Sallachadh**   32.283492   101.4624
33 **Dun Telve**   32.283492   101.4624

If these diameters are taken as 32.256ft, about one third of an inch short of the measured distance, there would be 28 Standard Canonical royal Egyptian feet of 1.152ft. Interestingly, the perimeter would be exactly 100 Standard Geographic Greek feet of 1.01367ft.

34 Achvarasdal Lodge   33.070894 103.9371

The solution to this circle is 28 feet of the Root Geographic Russian foot which would be equal to 33.04106ft — less than a half inch difference of the measured length.

35 Dun Boreraig   33.267745 104.55577

At a little over an inch too long on the diameter, it is proposed that this is the least accurate circle yet dealt with at around one part 400. The diameter may be 32 Standard Canonical common Greek feet of 1.0368ft, this would yield a perimeter of 100 Standard Geographic common Greek feet of 1.0427245ft. This is the outer diameter of the Stonehenge lintel ring.

36 Gurness   33.333362 104.76199

This is *exactly* 440 to 441 35 Root Reciprocal Roman feet at .952381ft.

Gurness, Orkney showing how the dwellings clustered within fortifications close to the broch. A sensible arrangement with Vikings for neighbours

37 Carn Laith   33.464596 105.17444

35 Root Reciprocal Roman feet, error 1:600 or 1/3 inch.

38 Clumlie   33.661446 105.79312

33.6ft would be about ¾ of an inch short of 35 Root Roman feet. (1:550).

39 Dun Osdale   33.858297 106.41179

This diameter is only about a tenth of an inch short of 35 Roman feet of .96768ft at 33.8688ft. (Standard Canonical). This length is also at an ideal 33.985 thirty five Roman feet of the Root Geographic .9710ft

40 Keiss West   33.98953 106.8242

This length is also at an ideal 33.985 thirty five Roman feet of the Root Geographic .9710ft

41 Forsinain   34.120764 107.23669

This circle is less than ¾ of an inch too long to be exactly 35 Standard Geographic Roman feet of .9732096ft.

42 Dun Beag   35.367484 111.15495

This circle is less than ½ inch short of being 35 Root Geographic Greek feet at 35.4ft.

43 Burray East   36.08927 113.42342

At 36.0818 feet this length is 35 Standard common Greek feet of 1.030909ft, about one tenth of an inch difference on the measured diameter.

44 Keiss South   38.320243 120.43505

There are two possible interpretations of this length; it could be 35 feet diameter of the root Sumerian foot of 1.097142ft, in which case it would be an error of slightly under an inch at 38.4ft. Alternatively at the same length, it could be 28 cubits of the Iberian Root value or 14 varas.

45 Torwoodlee   39.304496 123.52841

At 39.3346ft, about 1/3 of an inch difference, it is equal to 35 feet of the Root Geographic archaic English yard and full hand.

46 Thrumster   40.223132   126.41556

Almost exactly, at 40.22857ft, there would be 35 Root Canonical royal Egyptian feet.

47 Tirefuar   41.732319   131.15872

At 41.709ft this diameter would be 42 common Egyptian feet of .993071ft. It is also 40 common Greek feet of 1.04272ft. The former is the more likely interpretation, as the length would be 28 cubits or 7 fathoms of this foot.

48 Dun Ardtreck   44.488227   139.82014

This measurement is taken from the surviving semi-circle. At 44.4528ft, within less than ½ inch of the measured length, this is exactly 42 Standard Canonical Persian feet of 1.0564ft. This offers the same numerical interpretation as Tirefuar.

49 Oxtrow   44.816311   140.85126

At the value of an extended Persian foot times 42 at 44.808422 — this is less than 1/10th of an inch from the measured value; it offers the same numerical solution.

Although the above interpretations of the broch dimensions are the simplest, therefore the most likely solutions, within such a tightly related organisation of measure, alternative resolutions are possible. Site 1, Mousa for example, although this broch is seven "yards" in terms of the Assyrian foot it may also be viewed as seven steps, the 2½ ft module, whose detection in megalithic monuments gave rise to the belief in the "Megalithic Yard". At Mousa the step would be 2.5 Belgic feet of 1.08ft, therefore 2.7ft. Other values of this Belgic foot as well as variants of the Sumerian feet would yield a range of measures acceptably close to the hypothesised 2.72ft Megalithic Yard. For reasons that have become obvious, it is folly to attempt to define such a module as a habitually employed element of the megalith builders. If the seven division of no. 7, Keiss North, is relinquished for a division by eight, it would be eight Belgic yards whose constituent foot is 1.08ft, the perimeter would then be an integer in terms of modules the 175th part longer. This perimeter would then be 25 such *yards* composed of feet of 1.08617ft; as well as 25 *yards* this could also be expressed as 30 *steps* of 2.715428ft.

A typically commanding coastal setting of the brochs

It is also noted that not all of the diameters *can* be expressed in multiples of seven. Broch numbers 8, 9 and 35 may only be divided by multiples of eight. It is unclear why the seven counting base for diameters is sometimes abandoned; but it is often encountered in ancient metrology. Perhaps there was some compelling reason that a broch or circle had to be exactly a particular size, leaving but small choice as to the module.

Emendation note: *It is now known that Dunrobin Wood (no. 8) with diameter 26.378f t is 14 long cubits (a two feet measure) of the Root Samian feet. Broch 9, Brae, with a diameter at 26.5748ft has the same solution with a Samian variant the 175th part longer. Broch 35, Dun Boreraig, at 33.276ft has a diameter that is 14 2½ feet steps of the Standard Canonical Samian foot (the Root value × 1.008. Thus all of the brochs have the same solution in as much that their diameters produce a known module as the 14th division.*

There is also a distinct possibility that certain canonical lengths should be expressed in the constructions. Echoes of a far older metrological discipline are perpetuated in certain of the brochs. It had previously been noted that certain lengths seem to be equally comfortable when expressed as either a perimeter or a diameter in circular structures. The examples of such occurrences in the brochs are the *perimeters* of no. 25, Kintradwell and no. 35, Dun Boreraig; they are respectively the inner and outer *diameters* of the Stonehenge lintel circle. The evidence suggests that such metrological standardizations were common in the Iron Age.

One example being the wheel gauge of the chariot recently excavated in Yorkshire; remarkably, at 1.45 metres it is identical to that of the Edinburgh Iron Age chariot burial. This is the lesser value of the five Roman feet "pace", as found at broch 36, Gurness, the diameter of which would be 7 such five feet paces of the chariot gauge to 99.9% accuracy. This particular gauge is found in wheel ruts, whether they have been inadvertently or deliberately cut, throughout the ancient world. Notable examples occur in Pompeii, Malta, Corinth and in Persia.

Yorkshire Iron Age chariot in situ

Such observations as have been made here concerning the broch dimensions with regard to eliciting the rational sets of numbers, can be equally accurately applied to the older megalithic circles. As indeed they may be applied to interpret later cosmologies, such as the Saxon "King's Girth" or older biblical metrology such as the dimensions of the Mosaic Tabernacle. The Romans used the same criteria in founding their towns, as did Atribates whom they supplanted. Recent excavations at Silchester have revealed an Iron Age street grid that is in all respects similar to the imposed Roman, but angled at a 45 degrees slant to the Roman.

The statement that the system of measures has been accurately maintained from very remote antiquity until the present day is very easily demonstrated. As the best preserved megalithic ring in Britain is deep within the domain of the brochs, on Orkney, the Ring of Brodgar is as good a demonstration of this fact as could be wished for. Alexander Thom gave the diameter as 340.7 ± .44ft and stated that this was 125 Megalithic Yards. At 340.90909ft, exactly within the measured range, this would yield a Megalithic Yard of 2.727272 feet (Root Reciprocal); this is the vara as preserved in California and is 175 to 176 to the vara of Castile. At precisely 2.742857ft (Root) this was the official standard used by the Spanish bureaucracy until very recent times.

However, if one divides this diameter by seven, a more rational module emerges. One seventh of the Brodgar diameter is seen to be 10 five feet paces whose constituent foot is the Root Reciprocal value of the Common Egyptian foot of .97403ft, the perimeter is consequently 1100 common Egyptian feet or 220 paces. Even more obviously this perimeter, at 1071.428ft is exactly 1000 Root Belgic feet (found at broch no. 20, Kylescu) giving a closer pace to the human equivalent of 200 at 5.35714ft. It was the detection of this sort of module that led Alexander Thom to call it the Megalithic Fathom, which he tried to pin down to a constant of 5.44 feet. There is no such constant; each ring must be dealt with individually and its metrological solution sought in the rational numbers that emerge. It is of little use to try to divide diameters with a preconceived module, one must divide the diameter by one of the canonical *numbers*, and then see which module emerges (this cannot be done using metres, one would have to use a module of one third of a metre). The major oversight that prevented Thom from pinning the system down, was the fact that he showed no particular preference for his solutions to be in rational numbers of his proposed module. Near enough was good enough.

Emendation note: *The above interpretation of the Brodgar geometry has also been improved upon in the course of time. The reason that it was not seen at the time of writing was the reluctance to disagree with Alexander Thom's findings although the cracks in his Megalithic Yard theory were becoming glaringly apparent. With increasing familiarity with the universal metrological system his findings are now somewhat discredited but to express this at that time almost amounted to heresy.*

The error came about by slavishly accepting Thom's diameter as 125 Megalithic Yards and finding this proven module (2 1/2 Sumerian feet) within his fine survey parameters. I discussed the matter with John Michell and he proposed a more correct solution to the Brodgar geometry that fits equally accurately and far more logically. The surveyed diameter given as 340.7 ± .44ft is more properly 340.62336ft and this

both 14 times 24 Greek feet of 1.01376ft and 14 times 25 Roman feet of .9732096
feet and both of these values are repeatedly found in structures of all nations and
all ages. 24 feet is a double pertica and 25 feet is 5 paces. Both of these classification
values are found in the brochs, at Kintradwell and Dun Telve.

The fact of the matter is, that the vast majority of the megalithic rings can be metrologically interpreted by the methods that have been demonstrated on the brochs. All that is necessary is knowledge of ancient metrology, the foot lengths and their multiples as modules, which is nowadays universally lacking. Sadly, this is a development that has come about in the last half-century, it was not always so. Until the demise of Flinders Petrie the majority of archaeologists had a fair working knowledge of the subject. In the older editions of encyclopaedias such as the 1911 and 1915 editions of Britannica, Petrie wrote very extensive articles on the subject, in which he identified and listed over many pages, examples of all of the modules that we have discussed here. In modern editions scarcely a paragraph is devoted to the subject.

The broch builders therefore preserved methods and modules that had been used by the megalith builders that predated them by millennia and the same modules survived into the present epoch. Although the instruments of measurement may wear out, the standards by which they were manufactured would be accurately maintained in the dimensions of that which was already built. The conjectural purposes that are proposed for brochs, as well as being the residences of chieftains, council chambers, courts, temples or redoubts could also have been the local Weights and Measures bureau in its very dimensions.

The reason that we can now be certain about claims concerning metrology is that we are dealing with absolute values. No longer may the subject be regarded as arbitrary nor conjecture be utilised to substantiate hypotheses. One very good example of the solidity of the theory is the regularity with which the Assyrian variants occur in all cultures. Oppert positively identified the Root value of .9 English feet from measurements of the ruins of Assyrian Khorsabad. The value of the 175th part longer at .905142ft is exactly given by the copper bar of Nippur, at four feet long it is reported as 1.1035 metres and four times .905142ft is 1.103549 metres. At the next value in the series, the 175th part longer again, it is exactly the 360th part of the outer perimeter of the Stonehenge lintel ring. This particular value was precisely given by Stecchini taken from the diameter of the Grave Circle at Mycenae. These and other values of the Assyrian foot are also referred to as Oscan, Italic and Mycenaean. It therefore comes as no surprise to find it so prominently in the broch dimensions at Mousa, Midhowe, Borrowstone, Yarrows and Loch of Huxter at these identical values. Equally strong evidence is extant for each of the other proposed measurements.

Interior furnishing, broch of Gurness

Although Livio Stecchini, who has sadly died in recent years, was the most renowned metrologist of his generation, he too missed the fact that the choice of module must be sought in sensible ratios and rational numbers. His solutions often end in repetitive factions, and this is a clear indication that an alternative module should be sought. When he identified the "Mycenaen" foot from the grave circle of Mycenae he measured the diameter as exactly 100 feet of what I have termed the Assyrian foot at its Root Geographic classification of .910315ft. It has been my experience that only in certain cases is a decimal intentionally used as a diameter, and when this occurs is when one first looks for alternatives. The more likely scenario is that when this 100 feet diameter is divided by seven it is 13.0045ft, this is exactly, and more rationally, the 12 feet pertica of the Belgic foot of 1.083708ft. There is little doubt that we are looking at identical construction techniques and formulae over a vast geographic area and span of time.

Emendation note: It has since become obvious that 100 of a module is indeed often selected as a diameter. See "Measuring the Megaliths"

Few examples of ancient measuring instruments survived in Europe, and no ancient plans or diagrams remain; but it is obvious from the similarity of the broch designs that such detailed plans must have been used. It is to Egypt that we must look for pictorial confirmation of the facts regarding metrology as presented here. An abundance of measuring rods are extant and analyses of the dimensions of ancient buildings in very good condition may be used to confirm many modules. Many working drawings may also be consulted, a good example of which is set out below.

British Museum, wooden board overlaid with gesso. Egypt-5601

The human form is always depicted to canonical proportions, and the royal cubit length is taken from the sole of the foot to above the knee. The reason that the drawing here is so interesting is that seven different cubits are portrayed to the right of the figure. This is proof that an amalgam of modules may be deployed in a single scheme (let alone a single culture).
If the grid is in terms of the four-digit palm, and the cubit is taken as Root at 1.714285ft then the median of the cubits on the left would be this royal Egyptian cubit. The one above would be two Roman feet at the precise value found at broch 36, Gurness, and the one below would be the Sumerian cubit of 1.645714ft which is exactly 24 to 25 of the royal Egyptian, and found in broch 44, Keiss South. Only one cubit may be positively measured on the

right hand side, and at 1.571428ft it is directly related to the Persian values.

Were the figure erect, then the 20 four-digit grid length is 3⅓ Royal cubits, or more properly, exactly three of the two-foot Roman cubits. It is therefore obvious that although the draughtsmen reckoned in increments of the royal cubit, the finished artefact could be expressed as an integer in a related measure.

Many more of their metrological techniques may be extrapolated from this drawing, but the object here is to illustrate that several quite separate modules were in contemporary use in a single culture. Wherever one researches ancient measurement one finds the same modules and all of them are founded on such anthropometrical bases. There is nothing anomalous or even unexpected in finding the same system used in Scotland, after all, the modules are identical, the Root royal cubit used above would fit exactly 14 times into the diameter of broch 6, Dun Carloway.

Indeed, the common Greek foot is the basis of recent Scottish measure, namely the "elwand" which is a three feet measure of this foot it is found above in the brochs of Ousedale Burn, Dun A' Chaolais, Dunboreraig and Burray East; a two thousand year continuity, at least.

Virtually every aspect of this amazing and elegant system, particularly with respect to the module identification, would be obscured by being expressed in the metric system. As the traditional units such as the English foot are being inexorably phased out, we may confidently say that this knowledge has therefore been rescued in the nick of time. Only somebody who habitually thinks in terms of the English foot could have deciphered it. This is because each number that one is confronted with will have a close solution in terms of the English foot. For example, the length of the mean geographic degree calculated from the mean radius of the earth according to ancient metrology is 364,953.6 feet; the closest round sensible number to this is 360,000, when this degree is divided by this number the result is 1.01376 which is the value of the so called Olympic Greek foot commonly given as 309mm. (This is the foot in the perimeters of brochs 32, Sallachadh and 33, Dun Telve). Metrological analysis really is this simple.

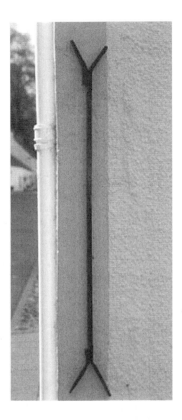

*A Scottish "Elwand" a three feet measure of the common Greek foot
this example is in the porch of Dunkeld Church*

It would appear that all cultures used all the measures. Although one cannot conjecture that there were any direct cultural contacts between the disparate peoples who used the identical system – does the system as a whole, which all civilizations used as reference, consequently predate all of them? Indeed, people who have not previously been regarded as civilized in the literal sense, manifestly utilised this sophisticated measurement system to extraordinary degrees of accuracy. But so thoroughly have assumptions over the centuries become orthodoxies, that the truth when it arises is often regarded as preposterous.

Structures from Sardinia, of the same period as the Scottish
Brochs, built by the Nuragic civilisation

Book IV

MEASURING THE MEGALITHS

AN ENQUIRY INTO THE MYTH OF THE MEGALITHIC YARD

Preamble - Studies Since 1997

This article was written as a booklet in 1997; it is a refinement and an enlargement of the previous *"Megalithic Measurement System?"* It is not simply a statement against the existence of the megalithic yard, but rather it is a cool as possible search *for* its existence. The conclusion was that there is no metrological evidence for its proof but there is good evidence of a unifying formulaic procedure in the overall plans of megalithic rings.

Having stated that, there has subsequently come to light a new major player in this arena whose general theory would point to the presence of both the megalithic yard and rod in the layouts of the Brittany alignments. This is Howard Crowhurst. His findings are set out in *"Carnac, The Alignments"* and he claims that a series of squares are conjoined in rows and that the hypotenuses of the resulting rectangles then form the sides of other smaller squares joined into other rectangles.

He argues very persuasively and produces the evidence of his claims. These layout rectangles would not be visible in the completed work, but appear to delineate the points of the ensuing geometry. Because he is dealing with the hypotenuses of non Pythagorean rectangles then the resultant "Megalithic yards" are not uniform; they conform to rough lengths that put them outside of the field of strictly defined unified unit fractions that govern metrology generally. This may shed light on Alexander Thom's otherwise unintelligible statement that the Megalithic yard was *"too good for the measuring stick."*

Regarding the presence of Thom's Megalithic yard as the construction unit of stone circles, one may state categorically that there is no evidence for its existence.

MEASURING THE MEGALITHS

4.1 THE EXISTENCE OF THE MEGALITHIC YARD

It is fifty years since Alexander Thom proposed that Megalithic circles were painstakingly constructed — as opposed to being roughly assembled. He explained their strict geometry by means of accurate site surveys and by the same methods demonstrated that the majority were astronomically aligned.

Had he stopped at that, there would be very little that is overly contentious concerning his claims. However, he also postulated that they had been laid out using a consistent unit of measurement throughout the wide geographic range wherein they are found. Fifty years later the existence of his proposed "Megalithic yard" is still an ugly bone of contention. Adopted by many as an established fact it has spawned countless poorly researched theories concerning ancient measurement.

Thom is directly responsible for this by stating in his *Megalithic Sites in Britain* chapter five: *"It is one of the objects of this chapter to demonstrate unequivocally the existence of a common unit of measurement throughout Megalithic Britain and to show that its value was accurately 2.72ft."*

Further to this: *"We first demonstrate that there is a presumption amounting to a certainty that a definite unit was used in setting out these rings. It is proposed to call this the Megalithic yard (MY) and two of these the Megalithic fathom."*

These are extraordinarily specific claims, particularly from such a qualified scholar, and seem to have been repeatedly – equally unequivocally – rejected by the many statisticians who have examined the evidence. Although this scientific rejection should have totally discredited the unit's existence, the hypothesis has proven hard to kill. At best the "Megalithic yard" is a metrological nuisance, and at worst has badly occluded the very beautiful structure of metrology. Having said that, it must be stated that the very proposition of a Megalithic yard has prompted a great many people, including myself, to study the *subject* of metrology in greater detail, helping to salvage it from the total obscurity into which it had sunk.

Although Thom had first proposed the MY as early as 1954, it first attracted wide attention in the late sixties, but the statisticians had seized upon it almost immediately.

The British statistician Simon R. Broadbent in the mid-1950s, using such techniques as lumped variance, carried out the first attempt at an analysis. He was followed by Professor D G Kendall, who used Fourier-analytic ideas on the data, two years later Peter Freeman reanalyzed the data from the different technique, that of Bayesian statistics. In summary, most professional archaeologists have never taken the Megalithic yard very seriously and the statistics have not firmly supported it either. None of the statisticians could present a firm case for the MY, all reported their findings as "inconclusive." This is not surprising, as the module that they sought is not there, at least, not in the form that was originally envisaged. The reason that neither Thom nor the statisticians were able to identify the module is that none of them had made any more than a cursory study of ancient metrology – or none at all. Therefore, the modules that did arise out of their statistics went unrecognised.

4.2 The Identification and Nature of the Module

Throughout the years of controversy, discussion and scientific analysis of the Megalithic yard it seems utterly incredible that not one person has been able to identify what it actually is. Its critics and detractors have come closer to the truth than have its supporters. This is because the main disparagement of the measure is that it resembles the human step and pace, and that is *exactly* what it is. This is not to say that a step or a pace was not a strictly defined standard; or that the Neolithic designers crudely paced out their ground plans, which is clearly their implication. In fact, the circles were laid out to greater degrees of accuracy than even Thom had proposed, but *never* in "Megalithic yards".

Thom identified far more elements than the yard and fathom. He discovered from his examination of cup and ring marks and their geometry that a subdivision of one fortieth of the megalithic yard had been used in their design; he termed this the Megalithic inch. He also proposed that a longer measure, that of 2½ Megalithic yards had been used in the perimeters; he termed this the Megalithic rod. There is enough information here, not only to define what the correct terminology for these measures is, but also to define from which branch of metrology they stem.

To anybody who has closely studied the structure of ancient metrology the nature of the Megalithic yard is immediately transparent. (This is to say that perhaps *nobody* has committed to this study). The Megalithic yard if it has forty sub divisions is clearly a *step* of 2½ sixteen-digit feet. Therefore the Megalithic inch is a *digit*, the Megalithic yard is a *step*, and the Megalithic fathom is a *pace*. The Megalithic rod is a little harder to classify, at 100 digits more divisions than 2½ steps are possible, it could be viewed as five twenty-digit *remens* or *pygons* and so forth. The structure of metrology has a limitless flexibility whilst adhering to strictly defined standards. The table below shows these measures with their Greek terminology.

MODULE NAME	DIGITS	FEET	MEGALITHIC EQUIVALENT
FINGER	1		Megalithic inch
Knuckle	2		
Palm	4		
Hand	5		
Lick	8		
HANDLENGTH	10		1/4 Megalithic yard (palmo)
Span	12		
FOOT	16	1	
Pygme	18	1.125	
PYGON (REMEN)	20	1.25	1/2 Megalithic yard
Cubit	24	1.5	
STEP	40	2.5	Megalithic yard
Xylon	72	4.5	
PACE	80	5	Megalithic fathom
Fathom	96	6	
(10 handlengths)	100	6.25	Megalithic rod
POLE	160	10	2 Megalithic fathoms

Note how the versatile digit structure allows for many counting bases, particularly sexagesimal, but the Megalithic is clearly the remen or groups of twenty — two-tens.

Thus, it is clearly demonstrable that the proposed Megalithic measures differ not one whit from other systems of antiquity, for all of them fall into this broad outline of the above structure. It is rather obvious that Thom had indeed hit upon values that were quite common in all ages. What Thom could not overcome, was the modern concept of what standards *ought* to be – this manifested the desire to find a single standard common to all the monuments. But the metrology of antiquity doesn't work like that.

Although he stated that the Megalithic yard was 2.72ft his surveys show a very perceptible variation from this value from circle to circle. Furthermore, they are often expressed in quite meaningless numbers like 23½ Megalithic yards. Surely reason would have it, that a rational design would employ rational numbers in the modules of construction? The modules that emerge from the surveys at approximately the MY value dsiplay very little constancy in this regard.

In identifying its branch, it is customary to reduce any module to its basic foot value in order to classify it. Therefore as a step, the MY of 2.72ft would be divided by 2½ to see which foot it conforms to; at 1.088ft the closest of the national feet to this value is the "Belgic" foot at 1.08864 feet making a Megalithic yard of 2.7216ft exactly. However, the structure of metrology allows for many other similar lengths to vie for the title of "Megalithic yard" well within the parameters used by Thom.

4.3 A Brief Explanation of the Arithmetical Structure of Metrology in General

The above accurate representation of the Belgic foot is stated in absolute terms and the ability to do this is a new phenomenon in the field of measurement. For in recent years more about the structure of metrology has been learned than at any comparable time in history.

Previously, the attitude toward ancient and historical social groups regarding measurement has been that it is arbitrary and inexact. It has always been known that there is a considerable variation within each of the national standards which has been interpreted as poor regard to their maintenance. However, it is easily demonstrated that *all* systems of measurement display the identical variations. Consequently, it is not unreasonable to assume that these differences are quite deliberate.

The two principal fractions by which measures vary are the 175th part and the 440th part. Although these fractions are very small, when they are multiplied into the dimensions of the monuments they are distinctly measurable. The 440th part of the side of the Parthenon is about 7 inches and in the side of the Great Pyramid is *exactly* one royal cubit. The reason for these variations in the standards is purely arithmetical, for they enable the principal outlines of any structure to be *expressed in integers*. Indeed, this would seem to be the raison d'etre of the structure of metrology.

The separation by these fractions does not happen once, but a series of times in each module, the following table shows all of the core variations

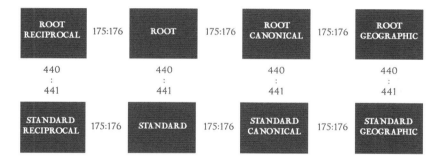

Each value across the rows increases by the 175th fraction and each value of the columns varies by the 440th fraction. The classification names have been assigned by their nature. The table below shows one of the foot measures — the English foot — placed in the "Root" position:

CLASSIFICATION TABLE

GREEK/ ENGLISH	Root Reciprocal	Root	Root Canonical	Root Geographic
	0.994318ft	1ft	1.005714ft	1.011461ft
	(30.307cm)	(30.479cm)	(30.654cm)	(30.829cm)
	Standard Reciprocal	Standard	Standard Canonical	Standard Geog.
	0.996578ft	1.002272ft	1.008ft	1.01376ft
	(30.376cm)	(30.549cm)	(30.724cm)	(30.899cm)

It has often been remarked that the English foot more closely resembles the Greek than the Roman. The above table shows that the English foot is one in a series of what has been termed "Greek" feet and there are numerous examples of all of these different values (of what are essentially the same foot) being used throughout the world.

As stated, the feet of all nations are subject to these listed variations. The national terminology, such as Greek, Roman, Egyptian, Persian, Belgic etc. that we assign to each of the modules means very little because all of them are but branches of a single organisation that was used concurrently in all nations. Let us first describe this integration in terms of what are widely acknowledged values of certain of the national feet. Livio Stecchini, probably the best-informed metrologist of the latter 20th century, noted many of these fractions linking ancient national standards:

Mycenaean / Italic foot	15	9		
Roman foot	16			24
Greek foot			10	25

The Roman foot is well known to be 24 to 25 of the Greek foot it is also 15 to 16 of the Mycenaean, the Mycenaean is also 9 to 10 of the Greek. By adding the Belgic foot to the list at the accepted 9 to 8 of the Roman, it is immediately discernible that the complex integration goes a bit further than Stecchini's reasoning – and that what we are looking at is a system that integrates by unit fractions.

Mycenaean	15	9	5		
Roman	16			24	8
Greek		10		25	
Belgic			6		9

So, the Belgic as well as being 8 to 9 Roman is also 6 to 5 of the Mycenaean, and the longer this national list is extended the more the integration increases in complexity but remains logical and straightforward.

If twelve of the national feet are considered (although there are rather more of the distinct feet than this) this will give a good example of the total integration of measurement. The table overleaf is a mere extension of the below list with the columns as connective fractions, Assyrian 63 to 64 of the Iberian and 15 to 16 of the Roman etc.

These twelve feet have values in terms of the English foot as:

ASSYRIAN .9ft

When cubits achieve a length of 1.8ft such as the Assyrian cubit they are divisible by two, instead of the 1½ ft division normally associated with the cubit length. Variations of this measure are distinctively known as Oscan, Italic and Mycenaean measure.

IBERIAN .9142857ft

This is the foot of 1/3rd of the Spanish vara, which survived as the standard of Spain from prehistory to the present.

ROMAN .96ft

Most who are interested in metrology would consider this value to be too short as a definition of the Roman foot, but examples survive as rulers very accurately at this length.

COMMON EGYPTIAN .979592ft

One of the better-known measures, being six sevenths of the royal Egyptian foot.

ENGLISH/GREEK 1ft

The English foot is one of the variations of what are accepted as Greek measure, variously called Olympian or Geographic.

COMMON GREEK 1.028571ft

This was a very widely used module recorded throughout Europe, it survived in England well beyond the reforms of Edward I in 1305. It is also the half sacred Jewish cubit upon which Newton pondered and Berriman referred to as cubit A.

PERSIAN 1.05ft

Half the Persian cubit of Darius the Great. Variations throughout the Middle East, North Africa and Europe. Survived as the Hashimi foot of the Arabian league and the pied de roi of the Franks.

Assyrian	63	15	9	7	6										
Iberian	64					20	14	8	5	4					
Roman		16				21					49	24	14	7	
Common Egyptian							15				50				48
Greek/English			10									25			49
Common Greek			8					9					15		
Persian				7											
Belgic															
Sumerian									6						8
English archaic															
Royal Egyptian									5						
Russian															

BELGIC   1.071428ft

Develops into the Drusian foot or foot of the Tungri. Detectable in many Megalithic monuments.

SUMERIAN   1.097142ft

Perhaps the most widely dispersed module throughout Europe, Asia and North Africa, commonly known as the Saxon / Northern foot.

ARCHAIC ENGLISH   1.111111ft

This is the foot of the 40 inch yard widely used in mediaeval England until suppressed by statute in 1439. It is the basis of Punic measure and variables are recorded in Greek statuary from Asia Minor.

ROYAL EGYPTIAN   1.142857ft

The most discussed and scrutinised historical measurement. Examples are plentiful.

RUSSIAN   1.166666ft

One half of the Russian arshin, one sixth of the sadzhen. Variants at one and one half of these feet as a cubit would be the Arabic black cubit, (which is also the Egyptian cubit of the Rawda Nilometer), and Vicentine and Palladian feet.

These are the values of the feet at Root classification and if any of them are placed in the "Root" position in the classification table they would all

20	6														
	35	20	14	9	7	6									
21	36					48	24	15	9						
		21				49				49	9				
			15				25			50	27	15			
								16			28	24			
				10								16	35	20	
7					8				10				25	36	48
						7					10			21	49

be subject to the regular multiplications of the 175th and 440th parts to achieve their final value.

In order for values to approximate the Megalithic yard, two of the above measures multiplied by 2½ come within Thom's parameters. The first is the greater Belgic foot whose spread of values is as so:

BELGIC (STEPS)	*Root Reciprocal*	*Root*	*Root Canonical*	*Root Geographic*
	1.610796ft	2.7ft	2.715428ft	2.730494ft
	(81.828cm)	(82.296cm)	(82.766cm)	(83.329cm)
	Standard Reciprocal	*Standard*	*Standard Canonical*	*Standard Geog.*
	2.690760ft	2.706136ft	2.7216ft	2.737152ft
	(82.014cm)	(82.483cm)	(82.954cm)	(83.428cm)

The second is the Sumerian step of 2½ Sumerian feet as so:
(These are also expressions of the Spanish vara of three Iberian feet).

SUMERIAN STEP (FT)	*Root Reciprocal*	*Root*	*Root Canonical*	*Root Geographic*
	2.727272	2.742855	2.758528	2.774291
	(0.83127m)	(0.83602m)	(0.84080m)	(0.84560m)
	Standard Reciprocal	*Standard*	*Standard Canonical*	*Standard Geog.*
	2.733469	2.749090	2.764798	2.780597
	(0.83316cm)	(0.83792cm)	(0.84271m)	(0.84752m)

Thom made comparisons with his value of the MY to these lengths of the Spanish vara, which is composed of three Iberian feet linked to Sumerian by a ratio of 5 to 6. Thus 3 Iberian feet (a yard) are an identical length to a 2½ feet (a step) of the Sumerian. The reason that we assume we are dealing with a Sumerian step and not an Iberian yard is because it allegedly has the 40 digits sub division.

The traditional structure of measurement systems reflects musical harmonics as shown in the table below and diagram opposite.

Interval	Cents	Name
1:1	0	Unison
32805:32768	2.0	Schisma
2048:2025	19.6	Diaschisma
81:80	21.5	Syntonic Comma
531441:524288	23.5	Pythagorean Comma
128:125	41.1	Diesis
25:24	70.7	Minor Diatonic Halftone
256:243	90.2	Leimma, Pythag. Halftone
135:128	92.2	Major Chroma
16:15	111.7	Major Diatonic Halftone
2187:2048	113.7	Apotome
27:25	133.2	Large Leimma
10:9	182.4	Minor Wholetone
9:8	203.9	Major Wholetone
8:7	231.2	Septimal Wholetone
7:6	266.9	Septimal Minor Third
32:27	294.1	Pythag. Minor Third
6:5	315.6	Perfect Minor Third
5:4	386.3	Perfect Major Third
81:64	407.8	Pythag. Major Third
4:3	498.0	Perfect Fourth
7:5	582.5	Septimal Tritone
45:32	590.2	Diatonic Tritone
729:512	611.7	Pythag. Tritone
3:2	702.0	Perfect Fifth
128:81	792.2	Pythag. Minor Sixth
8:5	813.7	Diatonic Minor Sixth
5:3	884.4	Perfect Major Sixth
27:16	905.9	Pythag. Major Sixth
7:4	968.8	Harmonic Seventh
16:9	996.1	Pythag. Minor Seventh
9:5	1017.6	Diatonic Minor Seventh

15:8	1088.3	Diatonic Major Seventh
243:128	1109.8	Pythag. Major Seventh
2:1	1200	Octave

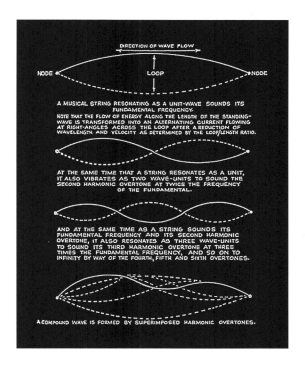

The diagram shows how naturally vibrating stringsdivide themselves as do metrological units; the 1st harmonic is a yard, 2nd harmonic is 2 cubits and the 3rd is 3 feet.

The following table of these harmonies is quite striking as illustrative of the precise correspondences with the system of metrology. This is because all of the decimally expressed harmonic ratios give a known module in terms of the English foot.

Up

F	4:3	G	3:2	1.1250000	3:4	0.5625000
G	3:2	A-	5:3	1.1111111	5:6	0.5555556
A-	5:6	B-	11:12	1.1000000	11:24	0.5500000
B-	11:12	C	1:1	1.0909091	1:2	0.5454545
C	1:1	D#	7:6	1.1666667	7:12	0.5833333
D#	7:6	F	4:3	1.1428571	2:3	0.5714286

Down

G	3:2	F	4:3	0.8888889	8:3	1.7777778
A-	5:3	G	3:2	0.9000000	6:2	1.8000000
B-	11:12	A-	5:6	0.9090909	10:6	1.8181818
C	1:1	B-	11:12	0.9166667	22:12	1.8333333
D#	7:6	C	1:1	0.8571429	2:1	1.7142857
F	4:3	D#	7:6	0.8750000	14:6	1.7500000

Harmonic ratios give identical values to modules expressed in English feet.

The first decimal column (*begins previous page*) reads the values of six of the feet that have been determined in this work.

1.125	is 9 to 8 of the English and is the foot of Nippur.
1.111111	is 10 to 9 of the English, an archaic English measure.
1.1	is 11 to 10 of the English, the Saxon or Northern foot
1.090909	is 12 to 11 of the English, it is the Sumerian foot
1.1428571	is 8 to 7 of the English foot, it is the royal Egyptian foot
1.166666	is 7 to 6 of the English, it is the Russian foot

The remaining six terms in this column (*the "Down" values*) are simply the reciprocals of the terms above. The first six terms in the second column are simply the half foot values. In the lower half of the second column only three of the terms can be stated to be modules of ancient metrology. 1.800000 is the Assyrian cubit, 1.71428 is the royal Egyptian cubit and 1.75 is the Russian cubit.

Although the other terms of the list are frequently encountered in metrological analysis there is no evidence that they were utilised as measures.

4.4 How One *Formula* Links Similar Constructions Not a Single Measure

The modern thought processes compel one to look for a single standard measure in buildings of a similar nature in the same culture. That is why modules based on the Roman foot are assumed to be the units of construction in Roman buildings, the Greek foot in Greek buildings and so on through all of the cultures. Experience shows us that this is not the case. In all instances there is a wide selection of modules being used, and in most cases several different modules are found in a single construction. Greek temples are patent in this respect because the procedures for determining these different modules are a matter of record, as in the writings of Vitruvius for example.

The Broch of Mousa, Shetland Islands. The best preserved of the Scottish brochs. Brochs were originally roofed.

The brochs of Scotland were inhabited round towers based upon quite accurate circles. They are Iron Age structures whose purposes remain a matter of conjecture. Were they redoubts, temples or manors? They may have been any or all of these, plus many more in their function. Many mysteries are attached to them, not the least of which is the presence of extensive vitrification of the stone in various of their number. Built from about 200 BC to 200 AD – they continued to be erected long after initial contacts with Imperial Rome.

They are firmly positioned within an area comparatively densely occupied by stone circles, some of which had been previously studied by the author. In the process of which, the simplest of techniques had been used in order to find their module of construction. The brochs, being circular, lent themselves ideally to the same method of analysis.

In 2003, whilst examining the dimensions of these Scottish brochs, it became apparent that although no constant measurement was detectable in their ground plans, there was unification in their conception. This observation came about through simply dividing their diameters by seven or fourteen. It is known through experience that integers were historically preserved in both diameter and perimeter by using 22/7 for pi. In every case a known module was detectable by dividing the diameter by 7.

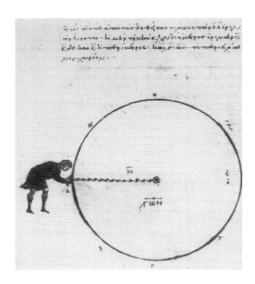

An illustration of Heron of Byzantium exemplifying the 22/7 pi ratio

Although there are over 200 broch sites, the sample of carefully measured inner diameters of the brochs that were available totalled 49; which is most certainly sufficient to establish what is going on with any potential unit of construction. The data is arranged in ascending order from the smallest broch at Mousa to largest at Oxtrow; they were respectively 18,9ft to 44.816ft.

Although this method invariably yielded a module, the surprising thing was that they were seldom the same. In other words there was no "broch foot" – and any attempt to find one is condemned to the same fruitless endeavour of trying to establish a singular "Megalithic yard" that rationally fits Megalithic monuments.

The first six brochs are sufficient to illustrate the foregoing statements because although they are fairly similar in their dimensions – from 18.9ft to 24ft – as the size of the brochs increase, the *numbers* of the modules do not. The modules themselves increase in order to maintain the numerical formulae. The first six brochs of the list illustrate this point:

Broch	Diameter	Module (closest)
MOUSA	18.9ft =	21 Assyrian feet of .9ft.
NYBSTER	21ft =	21 English feet.
OUSEDALE	21.84756ft =	21 com. Greek feet of 1.04036ft
CASTLE COLE	22.176ft =	21 Persian feet of 1.056ft.
ARMADALE	22.94ft =	21 Belgic feet of 1.092378ft
DUN CARLOWAY	24ft =	21 royal Egypt. feet of 1.14285ft

The remaining broch diameters all increased by a seven feet increment.

The following 20 brochs, from number 7 in the list, Kiess North, at 28.8934ft to number 27, Clachtoll, at 31.36ft, have diameters that are each of 28 feet which range from Iberian feet to archaic English feet.

From number 28, Midhowe, to number 31, Loch of Huxter, each are of 35 feet of the Assyrian variants; the diameters of the next three, from 32 to 34, revert to being 28ft of the greater measures, royal Egyptian and Russian.

From numbers 36 to 46 the diameters are again of 35ft in terms of the range of possible measures between the lesser Roman values ascending to the greater values of the royal Egyptian.

Finally, when the diameters exceed 40 whole English feet, the division of the final three brochs of the list, are in terms of 42 feet of the common Egyptian and the Persian standards.

Therefore each of the brochs was exactly seven of an accepted module. The first six are each seven *yards*, the following twenty are all a four feet module, called an *aune* in pre metric France, although widely used elsewhere there is no terminology for the measure apart from a double 2ft cubit. Then they increases to a *pace* of five feet and finally, the greatest of them, three in number, are *fathoms* of six feet, all of them in multiples of seven.

Applying the above techniques to the stone circles, identical results are obtained in module identification. This fact suggests that there is direct cultural continuity regarding math and measure spanning from the Neolithic to the end of the Roman occupation.

Dun Telve Broch, one of two brochs south of the village of Glenelg, Scotland

It would be difficult to assess just when in the Neolithic that stone circle building began. Certainly, the henges that often host them date from at least the fourth millennium BC and many dolmens are far older. Their flowering was most certainly the Bronze Age but the measurement system was there before they began and continued into our historical period

4.5 THE SHAPES OF THE CIRCLES AND THEIR DIMENSIONS

Previous to Alexander Thom, stone circles had been regarded as rather haphazard in their construction. By means of his accurate surveys he was able to establish that the not quite circular rings were deliberately so by demonstrating that their distortions were found repetitively. Moreover he showed that consistent geometrical methods had been used to create them.

In his book *Megalithic Rings* that he co-authored with Aubrey Burl, he illustrates the surveys of nearly 400 circles and distorted circles. He termed them Circles, and type A, type B, type D circles, Eggs and Ellipses. The true circles are far more numerous than these variants.

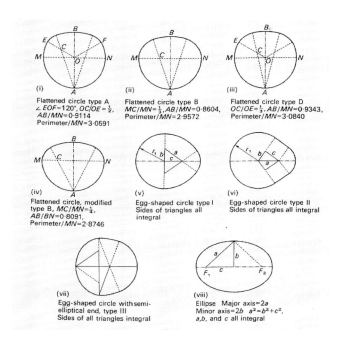

These are examples of the repetitive shapes of the underlying geometry that were found by Thom to govern both the stone circle and the cup and ring markings. Flattened circles, egg shapes, and ellipses. Picture taken from "Megalithic Sites in Britain and Brittany"

It is not proposed to use the shapes of eggs and ellipses to illustrate the metrology because these shapes are more contentious regarding clear values. The others at least have an obvious diameter to divide by traditional formulae. That is the types A, B and D are hemispherical on one side and this gives one clear radius as the datum. It is very easily demonstrated that a seventh division of these radii invariably gives a recognisible module. Whereas it is very clear that nothing approaching an approximation to the 2.72ft Megalithic yard is found in rational numbers.

However, as stated, if the accurate values of what the Megalithic yard actually is are used one *sometimes* finds totally accurate examples of the

module. Namely steps or paces in terms of the Belgic or Sumerian feet
Shown below is the "type A" ring of Black Marsh in Shropshire is given by
Thom as 76ft in diameter which he states to be 27.9 Megalithic yards. This
begs the question why on earth would someone wish a datum to be this
odd number? The correct solution is easily found by simply dividing the
diameter by 14, in this case the "so called Megalithic fathom" is a *pace* of
5.430857ft and this is the Root Canonical value of 5 Belgic feet to within
less than half an inch of the whole measured diameter of 76ft. (1.08ft ×
176/175 × 2.5 = Megalithic yard of 2.71542ft = diameter 28 MY).

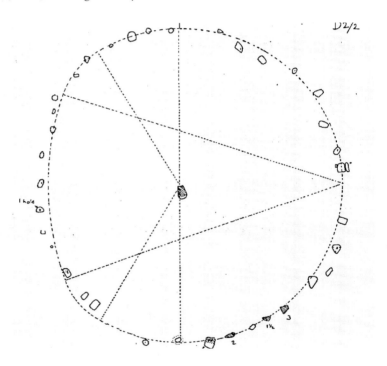

Black Marsh, Shropshire. P 27 Megalithic Rings

Only two other rings give a solution in near Megalithic yards to the
above formula of the 14 diameter division. One is Y Pigwyn in Wales at
28 Belgic steps of 2.730925ft each foot of 1.09237ft, the other is the outer
of the concentric rings of the Aquorthies Kingousie at 28 steps of each
foot of 1.071428ft at 2.67857. *Nowhere does any circle have a sensible integer
diameter in terms of a Megalithic yard of 2.72ft.*

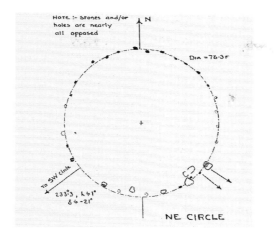

Y Pigwn stone circle, Powys p.391 Megalithic Rings

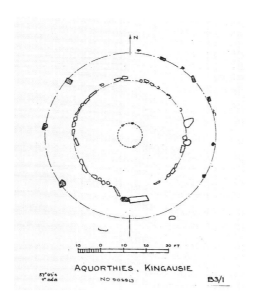

Aquorthis stone circle, Perth p.277 Megalithic Rings

In order to show the regularity with which the geometry of rings was exactly duplicated, Thom took two class A circles, Castle Rigg and Burnmoor in Cumbria and superimposed one upon the other. He stated that they only differed in scale and orientation (*see overleaf*).

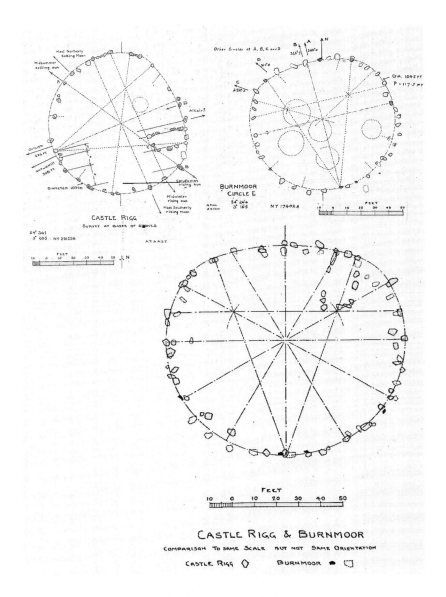

Above is Thom's illustration of this composite.

The difference in scale, as can be seen, is in the region of about 10 feet. Although a module may be isolated by the fourteenth division of these diameters, these particular circle's diameters are more interesting in that they are both very accurately modules of 100 feet. This practice is widespread in Megalithic circle construction, as we shall later see in the Stonehenge examination.

The Castle Rigg diameter is a perfect illustration of this fact as the module in question is identical to the value deduced from the Aquorthies Kingousie ring, previously mentioned.

The basic step module at Aquorthies Kingousie stems from the lesser Root Belgic foot of 1.07142ft and the diameter of Castle Rigg is 100 of these feet to within about ½ an inch overall. This is a far greater level of accuracy than one part in 400 that Thom claimed the Megalithic builders were capable of.

Similarly the Burnmoor circle has a diameter of 100 Russian feet of the Root Canonical classification ($1.166666 \times 176/175$) or 1.173333ft: this is an error of less than an inch in the overall circle from the calculated value. Listed below in ascending order are some of the other circles that have one hundred feet diameters:

KINGSTON RUSSEL, Dorset   91.1ft = 100 Root Geographic Assyrian ft = 91.238ft

TARMOOR, Westmoorland   96ft This is 100 Root Roman feet.

LILBURN, Northumbria   100ft = Root Greek foot

AULDGIRTH, Dumfries   100.2ft = 100.2727ft Standard Greek feet.

THE HURLERS south circle   102ft 100 lesser Root Common Gk = 102.04ft

AULDGIRTH,   102ft a/a

At this juncture we could interpose **ROUGH TOR**, Cornwall whose diameter at 150.7ft is 100 cubits and at Root Canonical Greek would be 150.8ft. (*see Forse circle, below*)

ROLLRIGHT, Oxford   103.6ft 100 Standard Canonical common Greek feet of 1.0368ft, within an inch of the measurement.

BURNMOOR E Cumbria   104.5ft.   100 Standard Geographic common Greek feet are 104.27245ft, accurate to about 2½ inches. This is also the precise diameter of the outer lintel ring of **STONEHENGE**.

GREY WETHERS, Devon   104.5ft a/a

TREZIBBET, Cornwall   106.2ft This is 100 Root Geographic Persian feet and Forse circle is 157.5ft which is 100 cubits of the Root Persian foot of 1.05ft exactly.

TRESWIGGER, Cornwall   108.3ft

SHELDON, Aberdeen   108.4ft. This circle is the outer of concentric rings.

It is assumed that these last two circles are meant to be the same (Standard Belgic of 108.2454ft). Should you wish these could also be viewed as 40 Megalithic yards of 2.706ft. Thom states the former as 125 Megalithic yards perimeter and the latter as 39.8 diameter.

URQUART   110ft is exactly 100 Root Saxon feet (Sumerian).

ELVA PLAIN   113ft 100 Root Canonical Nippur ft are 113.142ft.

FARR WEST, outer circle   113.2ft a/a

THE HURLERS north circle   113.6ft 100 Standard Canonical Nippur are 113.4ft

This list of 100ft diameters is by no means complete, but enough have been noted to establish the fact that the choice of this module length is a very deliberate custom. This is important to establish before the next chapter is commenced.

In all cases there is no measure that fits any of the circles in integral Megalithic yards of 2.72 feet. Yet diameters are invariably rendered into rational numbers in terms of other ancient modules, but *no single measurement* may be shown to fit all of them. There can be no Megalithic yard on these grounds alone.

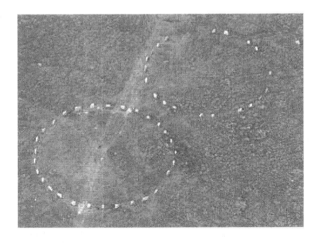

The Grey Wethers stone circles, on the edge of Fernworthy Forest, Dartmoor. Two circles remain, both of 104.5 feet, the same as the outer lintel ring of Stonehenge.

4.6 Stonehenge

The foremost institutional centre of Britain that governed the social structure and annual rituals was the temple of Stonehenge. Overshadowing its vast surrounding necropolis it differs so much from any other Megalithic circle that doubts have been expressed that it belongs to the same culture. The metrology proves that it does and so does its precise siting within the landscape with its undoubted geographic location related to other contemporary monuments such as Avebury and Woodhenge.

Throughout the world metrology has always been associated with the Temple; the dimensions of which are expressed in canonical numbers enshrined in the modules. Its fabric is the permanent repository of measures and it was the custodians who often manufactured and issued the instruments of metrology, both of weights and measures.

The metrology of Stonehenge is rarely discussed at length, its astronomical attributes receive far greater coverage, yet the genius of its numerical conception remains unsung. It is a peerless wonder in this respect and has taken many years to finally become interpretable – and when you believe the process is complete it never fails to surprise with yet more. It is a great yardstick of comparative metrology and it is really the Waterloo of the Megalithic yard for it is here that Thom's theory of a constant unit is most easily discredited. Neither the Megalithic rod nor the Megalithic yard may be demonstrated in any of the Henge dimensions.

Thom claims that there are 45 Megalithic rods in the inner sarsen perimeter and 48 in the outer sarsen perimeter. This would imply a lintel stone width of 3.247ft which every survey shows is far too short. The lintel width is usually given in the region of 3½ft and by the following reckoning may be expressed as the absolute 3.4757485ft.

The only diameter given by Thom is the inner of the sarsen circle which he states is 35.81 MY, therefore by inference the outer would be 38.2 which is 103.9ft. However if the true diameters are viewed in their correct modules the solution is far more elegant than these clumsy and quite meaningless numbers. The inner diameter, as established by Petrie, is exactly 100 Roman feet of .9732096ft and the outer is exactly 100 common Greek feet of 1.0427245ft.

Logic, as well as observation, dictates the monument's dimensions, for there are 30 uprights, 30 lintels and the width of the lintel is one 30th of the overall diameter at 3.4757485ft.

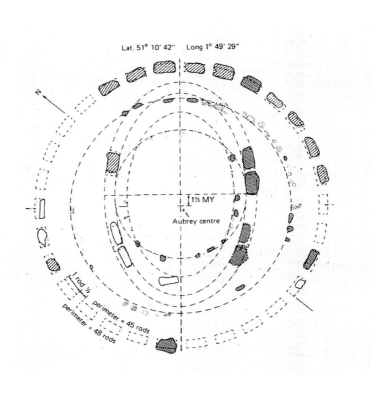

Stonehenge. P122 Megalithic Rings

Because of the integration of metrology these are some of the basic feet which fit these Stonehenge diameters:

The inner diameter

100 **Roman** feet of		.9732096ft
98 **Common Egyptian** feet of		.993071ft
96 **Greek** feet of		1.01376ft
84 **Royal Egyptian** feet of		1.158583ft

The outer diameter

105 **Common Egyptian** feet of		.993071ft

100 **Common Greek** feet of	1.0427245ft
96 **Lesser Belgic** feet of	1.086171ft
90 **Royal Egyptian** feet of	1.158583ft

In order to understand this integration these modules, being of Standard Geographic classification, should be divided by the formula of 1.01376 to see their true fractional relationship with the English foot at "Root".

The **Roman** foot is	.96ft	or 24 to 25;
the **Common Egyptian** is	.97959ft	or 48 to 49;
the **Common Greek** is	1.02857	or 36 to 35;
the **Lesser Belgic** is	1.07142	or 15 to 14;
the **Royal Egyptian** is	1.142857	or 8 to 7.

Thom has also noted on the above plan that the Aubrey holes centre is spaced at 1.5 MY from the sarsen centre along the sunrise axis of the monument. Doubts may be expressed at this conclusion because there is much play in the intended centres of the Aubrey holes and it is the author's belief that the intended dimension of the Aubrey holes can be determined through geometric and metrological methods with total accuracy by the following reasoning.

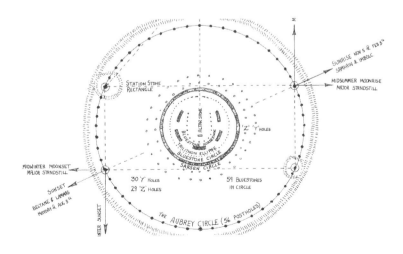

The Station Stone Rectangle at Stonehenge, illustrating the precise two Belgic feet cubit clearance between it and the sarsen circle.

I was prompted to look into the matter by the supposition of John Michell that the station stone rectangle, which is one of the phase 1 constructions and placed there a millennium before the sarsen stones, was in fact Pythagorean. It is five by twelve on the sides and has a thirteen diameter.

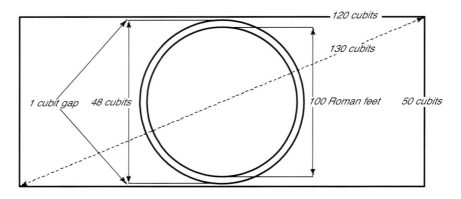

The Station Stone Rectangle showing how the proportion dictates the module

From their very close measurements of these stations I was able to identify the module that was used to lay it out as a two feet (long) cubit of the above mentioned Belgic foot of 1.086171ft (at 96 to the outer lintel diameter). The rectangle is 50 × 120 × 130 hypotenuse in terms of this cubit.

The conclusion was strengthened by the fact that if the sarsen circle was placed at the centre of this rectangle then there would be a space of exactly one cubit at either side of the sarsen stones and the side of the rectangle. As the station stones are aligned to maximum and minimum lunar standstills this 2 Belgic feet gap would provide a perfect squint between the stations and sarsen ring to observe this phenomenon. In metrological terms, this arrangement is too good to disturb, however this hypotenuse does not quite define the Aubrey holes, as it is reputed to do, because it is too short. Further doubts were expressed as many people believed that this rectangle defines the rectangle at the centre of an octagon. The final solution satisfies all points of view in the following manner.

It has always been assumed that the hypotenuse of this rectangle at 282.40457ft was the diameter of the Aubrey holes and that the station stones defined the Aubrey circle. I realised that this was not necessarily the case, because the dimensions for the Aubrey circle taken from this measurement made no metrological sense. This is the point that is totally missed by those who comment on megalithic circles – *the modules must exactly express rational numbers that are pertinent to the structure.* In the matter of the Aubrey resolution everything instantly came into sharp focus immediately this dimension was established.

Reliable estimates for the perimeter of the Aubrey is 271.6m. Naturally, we won't recognize anything from this, so convert to feet. It is 891.0768ft, as any perimeter will be a multiple of 22 the answer is rather staring at you; the closest multiple to the total is 880. Therefore you divide by that and it is 1.012587 which is acceptably close, at .12 of a percentage point, to the Standard Geographic Greek foot of 1.01376ft. If the estimated perimeter is adjusted by this fraction (it amounts to about a foot in the total length) this an entirely logical procedure, because all other measures of Stonehenge conform to this 1.01376 formula.

The diameter may therefore be calculated as 283.8528ft which is exactly 1.4482285ft longer than the station stone hypotenuse of 282.404571ft. This proves to be a module, it is exactly a Roman cubit of the lesser Standard Geographic classification. There is evidence that the same conclusion had been reached by the 1972 survey – that the station stones lay about half a foot within the mean of the Aubrey circle (it is 8.689ins). This Roman cubit then governs the geometry as so:

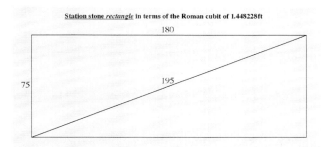

Station stone *rectangle* in terms of the Roman cubit of 1.448228ft

As shown, it is also Pythagorean, having a relationship with the Belgic numbers as 1 to 1.5.

If the rectangle were the innards of an octagon, as it has often been theorised to be, then it would have to be lengthened by exactly one cubit on the hypotenuse which destroys the integrity of number and appears as so:

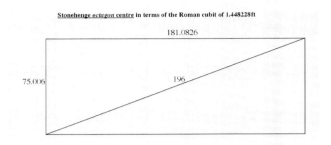

Stonehenge *octagon* centre in terms of the Roman cubit of 1.448228ft

181.0826

75.006

196

As can be seen, although the longer side is 1.5678ft longer and the hypotenuse is a full 1.448228 feet (one cubit) longer than the station stone rectangle, it *almost* keeps its integrity at 75 cubits to be the side of an octagon upon the Aubrey holes (remember, we are taking as our datum the diameter of the Aubrey holes). It is about a tenth of an inch too long. Does that disqualify it? Perhaps, but these are the levels of accuracy that we are talking about.

If one accepts the octagon then it would appear as so:

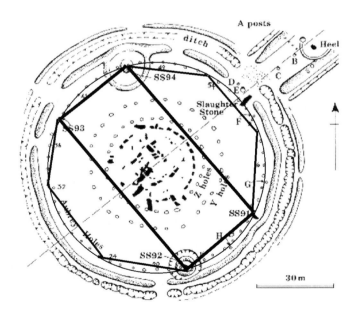

As there are 56 Aubrey holes divisible by both seven and eight then their distance apart upon the perimeter is exactly 11 cubits, the points of the octagon

upon the perimeter would therefore be 77 cubits apart.

Equally often it has been hypothesised that heptagonal geometry governs both the Aubrey circle and the dimensions of the sarsen circle. In heptagonal geometry the relationship between the overall radius and the radius of the connecting ring of the sides *intersections* is 1:2.8017197. It would appear that (because of the love of integers) this was rounded in these calculations to be 1:2.8, this makes a difference of only 1 and 1/22 of an inch on the entire 141.93ft radius; (for the sake of calculation the same was done to the pi value by accepting 22/7). This small adjustment brings the clarity to the metrology.

It is established that the lintels are accurately 3.4757485 feet then by using this formula it may be calculated that the mark would fall upon the lintel stones, not on the their mean or halfway point, but on a formulaic ratio that is dictated by the modules as identified above.

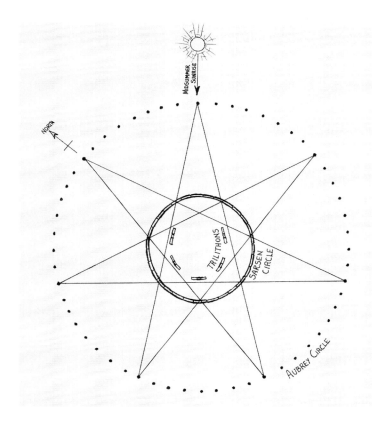

The Aubrey circle perimeter is both 440 Greek "long" cubits of 2.0725ft and 616 Roman 1½ feet cubits of 1.4482285ft. These combined measures are exactly the width of the lintel stone which is also three Royal Egyptian feet of 1.1585828ft. The inner and outer diameters of the lintel are respectively 100 Roman feet and 100 common Greek feet and the median dictated by the heptagram is 100 Greek feet, all are of the Standard Geographic classification.

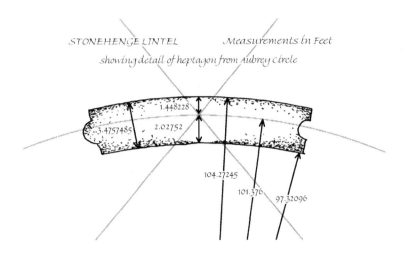

The divisions of the lintel circle at Stonehenge
as dictated by the intersection points of the heptagram.

Once again, to see what is happening here in straightforward mathematical terms all of these numbers should be divided by 1.01376 (reduced to Root) as was done with the original foot module above.

$$3.4757485 = 24/7$$
$$1.4482285 = 10/7$$
$$2.02752 \quad = 2$$
$$104.27245 = 100 \times 36/35$$
$$101.376 \quad = 100$$
$$97.32096 = 96$$

Strictly speaking, both heptagram and octagon are only theoretically present and it is my belief that this accounts for the nature of the Aubrey holes. The width of the Aubrey holes allows for these alternative measures and I forward the hypothesis that this is their purpose.

It is believed that they date from phase 1 of the monument, about 3100 BC. This early date is when the outer ditch was dug and the earth bank thrown up. The Aubrey holes were dug at the same time and because of the lack of weathering in their interiors, some archaeologists believe that they were immediately back-filled and they had never housed posts. It would not be unreasonable to propose that they were dug in order to sift out any hard objects, such as pieces of flint that could deflect any surveying pegs from carefully calculated positions when being driven into the earth. Long bone pins were found in many of the holes that also held cremated remains.

As in excess of 20 holes remain unexcavated this proposal could be investigated at some future date. If little or no hard objects such as one might expect find – (from neighbouring test holes) – are unearthed then their existence as the earliest layout that served to dictate all of the subsequent geometry would be supported. As there were several construction phases at Stonehenge until its abandonment around 1600 BC, it appears that all of the developments adhered to a single canon using the same carefully preserved set of measurements over millennia.

As has been illustrated, an understanding of metrology adds a whole new and previously neglected dimension to the study of monuments.

4.7 Identical modules from non Megalithic sources

When "greater" or "lesser" terminology is applied to a module it means that the "Root" has been altered by a factor of 1.008, many modules are subject to this change. Notably the Roman and the Belgic; the Roman .96 moves from the Root to the Standard Canonical position where it is more comfortable as a canonical number. By its reduction at Root to .952381 instead of being 24 to 25 of the English, it is 20 to 21. Similarly the Belgic Root of 1.08 is more suited as a Standard Canonical and its lesser Root at 1.071428 is then 15 to 14 of the English foot.

Here at Stonehenge the outer measures, Aubrey holes and rectangles, are largely developed from lesser Standard Geographic measures and the inner sarsen circle are from the greater Standard Geographic measures, all are governed by the formula 1.01376. This *number* of course, as a measurement in English feet, is the value of the geographic foot at the latitude of the 51st parallel where all this takes place. It may be regarded as a musical change and there is a delightful harmony between them.

In Stonehenge the greater and lesser Roman foot are both present, the inner sarsen is 100 greater Roman feet of .9732096ft and the foot of the cubit that is generated by the rectangle is the lesser at .965485ft. The separation is invariably 125 to 126 or 1.008. The first time that I encountered the lesser variant of the Roman foot at Root value was in the publication *"Angkor Wat: Time, Space, and Kingship"* by Eleanor Mannika of Michigan University and is the cubit that she had deduced from the Angkor Wat temple. Her value was .43545mm and the absolute value is .43543mm and you see it in the lintel stone picture on the previous page. In feet it is exactly 10/7ft.

Closer to home is the example of a Roman foot on the Greek Ashmolean relief. The length of the Roman foot that is above the right arm is .965485ft therefore "lesser" or the value we found in Stonehenge that is the basis of the cubit of the Aubrey holes.

"Vitruvian" Man in the Ashmolean Museum, Oxford

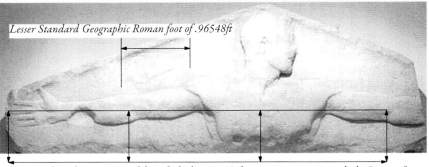

Lesser Standard Geographic Roman foot of .96548ft

Three cubits, the proportional foot of which is 1.1264ft, a seven to six ratio with the Roman foot
This is a Standard Geographic Archaic English foot of the "yard and the full hand"

Another famous building that was designed by the Standard Geographic classification modules is the Parthenon which has the precedent of having the modules identified for you. This temple to Athena was known as the Hecatompedon which means "hundred feet measure" and the stylobate, or platform upon which the columns are erected, is a width one hundred feet of 1.01376 feet. As well as this hundred footer another hundred feet measure is the length of the most important of the inner chambers, the naos or cella, wherein stood the great statue of Athena. It is 100 Roman feet of .9732096ft, exactly the inner diameter of Stonehenge. [1]

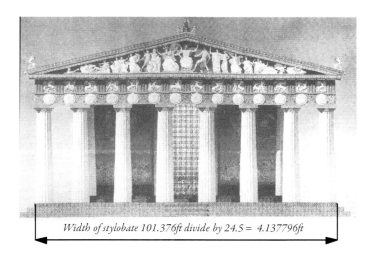

Width of stylobate 101.376ft divide by 24.5 = 4.137796ft

Vitruvius states: *"the stylobate of an octastyle building should be divided by 24½ to find the module"*. Thus you can see that these measures are very precisely identifiable and are as well preserved on Salisbury Plain as in ancient Greece. Vitruvius states the method of module identification in very plain terms and it is a great mystery that his methods have never been applied to this purpose. Nowadays this is because it is expressed in metres and the value of the module of the stylobate at 1.2612 metres means very little; but at 4.137796ft it simply requires dividing by the prefixing 4 in order to identify the module.

Like Stonehenge the guiding number at the Parthenon is 1.01376. You then divide the resultant 1.034449ft by this and it gives the Root, in this case 50/49 English feet. This is immediately classifiable as a *lesser* Root common

1 Penrose's measurement of the *naos* although he stated it was 100 Roman feet, did not take certain features into consideration; his Roman foot is therefore slightly off. Stecchini claims his given length should be reduced by 23.5cm (ca. 9 inches) according the measurement of Lucien Magne who was commissioned to assist with Parthenon restorations. (*American Journal of Archaeology*, 1905 Vol 10.) His resultant length is within 5mm of 97.32096ft.

Greek foot. (The seven ashlars of the architrave are each (ideally) 14 of these feet of 1.034449ft). The value in Stonehenge is of the greater Root i.e. 1.008 greater than the Parthenon at 1.0427245ft and this is the 100th of the lintel diameter. These modules are regularly found by archaeologists and mathematicians *but they do not recognize them.*

As the final example of the variety of modules being used in a single structure here is a table of measures that the architect, John James, found to have been used in Chartres cathedral. This is the identification of the modules:

Module in mm	*converted to feet*
279.6 is **Iberian** the absolute of which is	.91639ft
282.5 is also **Iberian** with an absolute value of	.92686ft
285 is 6 to 7 of **Sumerian** measure, its absolute is	.93506ft
294.4 is a **Roman** whose absolute would be	**.96548ft**
305.7 is known to us as an **English/Greek** absolute of	1.0022ft
322.9 the **pied de roi**, same as "**Persian**" measure	1.0583ft
325.8 also **Persian**	1.0668ft
337.084 which is **Sumerian** and v. close to the absolute	1.1059ft
353.3 is a **royal Egyptian** foot whose absolute value is	**1.1585ft**

The two values that we have independently deduced from the Stonehenge dimensions, the Roman and the royal Egyptian, are those that are emboldened.

We could continue through every culture to infinity. For example the unit that was widely used in the construction of Teotehuacan in Mexico, as identified by Hugh Harlesden, is 3.47574ft or the precise width of the Stonehenge lintels, one 6 millionth of the polar radius. These ancient modules are truly universal and are at the foundation of every culture in their most minute detail.

A further and very pertinent example as recorded by Stecchini is that of the Grave Circle of Mycenae (foreground below). By his own measure, he stated it to be 100 Mycenaen feet of .9103ft in diameter. The Mycenaen is nine to ten of the Root Geographic classification of the Greek foot (1.011461ft). There would be exactly the highly significant perimeter number of 360 of these Mycenaean feet in the outer lintel perimeter of Stonehenge. *(This "Mycenaean" foot would be termed Root Geographic Assyrian).*

The grave circle of Mycenae, foreground

Just as these better known buildings are clearly shown to have many modules used in their construction so it may equally be shown that the same rule applies to the megalithic circles. To look for a single module employed in a great number of similar structures is absolute folly. All that is needed for such fantasies that waste everybody's time and energy to be abolished, is for those who comment upon ancient metrology to first verse themselves in its principles. They are essentially quite straightforward.

John Neal, Glastonbury, May 2007.

4.8 ADDENDUM CONCERNING NEOLITHIC TIMBER RINGS 2016

It was thought necessary to include these additions to *Measuring the Megaliths,* as these notes are pertinent to the scheme in the light of further information. The present author through ever widening researches into metrology had, quite early on, reached the conclusion that there is no such thing as a Megalithic yard *as Thom envisioned it.*

Its acceptance by a broad spectrum of qualified men had largely come about through Thom's impressive credentials, this is borne out by Dr Euan MacKie's positive opinion on the subject. It was recorded by Richard Dibon-Smith of the University of Toronto, in his article *The Search for Megalithic Quanta* ca. 2002; he stated: "One of the more enthusiastic appraisers of Thom's conclusions was Dr Euan MacKie, a professional archaeologist, who accepted the existence of the MY outright since, *"considering the qualifications of the author, [it] is scarcely likely to be challenged."* " In point of fact, the Megalithic yard hypothesis has been repeatedly challenged by both statisticians and archaeologists from the very beginning, which was Thom's introduction of the subject via a paper in 1955, followed by *Megalithic Sites in Britain* in 1967.

The archaeological community was fairly evenly divided in either its opposition or support for the theories of Thom. Surprisingly, the principal hostility to his Megalithic yard came from those who were closest to him in his research projects, particularly regarding his surveys. Aubrey Burl, for example, coauthored *Megalithic Rings*[1] with him in 1980; this book was simply all of the illustrated survey plans that were described in summary by each man in turn. Tellingly, although Thom often gave the results exclusively in terms of the Megalithic yard, Burl seldom mentions it. His opinion of it was made abundantly clear in his book *Rings of Stone: The Prehistoric Stone Circles of Britain and Ireland* in which Aubrey calls the megalithic yard *"....a chimera, a grotesque statistical misconception."* This was published in 1979; he had therefore reached his conclusion before coauthoring with Thom.

Aubrey Burl, being an acknowledged specialist of the European

---

1 Thom, Alexander, Thom Archibald Stevenson, Burl, Aubrey. "Megalithic rings: plans and data for 229 monuments in Britain", *British Archaeological Reports*, 1980.

Neolithic-Bronze Age, had the advantage in matters of metrology as applied to megalithic remains; he had taken an independent interest in the underlying designs and had formulated his own opinion on units of construction. Two sites in particular were singled out by Burl as exemplary of a constant unit; these were Woodhenge and The Sanctuary at Overton hill that is part of the Avebury complex of megaliths. Both sites are ideal to exercise metrological detection, because they are composed of concentric rings of both stone and timber posts.

Burl discounted the theory of the megalithic yard in these monuments, although he was not aware of the fact, he had sided with the interpretation of Maud Cunnington.[2] This is because he had isolated a unit of measurement that he stated had been consistently used in stone circles and passage graves both in Britain and Ireland; he termed the unit a *"beaker yard,"* named from the Beaker people. This was a widely scattered culture in Neolithic Europe from ca. 2500 BC. They were so-named from the inclusion of beaker shaped earthenware urns in their interments and these remains are often associated with megalithic monuments and barrows.

4.9 WOODHENGE

It was Maud Cunnington who had excavated both Woodhenge and the Sanctuary. Woodhenge was found quite by chance as the ancient concentric rings were spotted as crop marks from the air in 1925. The Sanctuary, on the other hand, she had deliberately sought out; it had been completely obliterated during the wholesale vandalism of stone monuments that were cannibalized for utilitarian usage in the Eighteenth Century. Fortunately, the antiquarians William Stukeley and John Aubrey had recorded its location and description.

Neither Burl nor Cunnington were particularly good at module identification, although both were very sharp at constructional quanta recognition. What this means, is that they detected (unlike Thom) certain

---

2   Maud Cunnington, married to Edward B H Cunnington, a fourth generation Wiltshire archaeologist. They purchased both Woodhenge and The Sanctuary from private ownership and bequeathed them to the nation.

modules in the course of their excavations through the straightforward logic of the rational numbers occurring in the design intervals, but failed to identify them. Alexander Thom recorded all of the dimensions of Woodhenge in terms of the megalithic yard, and although he expresses the dimensions in whole numbers, he uses sleight of hand to arrive at his figures. Woodhenge is an excellent example of this trait, because he purports to have found that all of the perimeters of the elliptical rings were an integer progression. This is an absurd approach, for the simple reason that it is in the axes of ellipses where the quanta should be sought, in the same way one begins with the radius of a circle as this determines all else.

The module derived by Cunnington from measurement of the long axes of Woodhenge she stated was 11½ inches. From the axes of the four inner rings she obtained this value in near integers of 40.1, 60.1, 80.2 and 100.2; considering that these all vary from an integer by virtually the identical excess, one wonders why she did not let these fundamental measurements dictate her module and alter it accordingly. Because had she done so, she would have exactly identified the Root Roman foot of .96ft, which would have corrected her module of 11.5 inches to be 11.52 inches. She did not do this because she would not have recognized it as a Roman foot, neither would she have seen the significance of it being exactly 24 to 25 of the English foot – the acceptance of the Roman foot generally being considerably longer, most often at 11.652 inches. Two conclusions may be drawn from these observations, the first being – that these monuments had been designed and executed to levels of accuracy that were unsuspected; the second is that having identified a module and a classification term, Root in this case, then any other modules that may arise from the Woodhenge geometry will be of this same classification.

Alexander Thom drafted the illustration of Woodhenge, reproduced below; it is the four inner ellipsoid axes that Maud Cunnington recorded as the near integer the series described above. Aubrey Burl had a different interpretation.[3] He stated that there were multiples of four in the number of posts in the six rings; 12, 18, 18, 16, 32 and 60. (18?) Additionally, the unit of measurement that he had identified elsewhere and had termed a "Beaker Yard" from its obvious associations, at 73cm it fit the six all-important

---

3 Burl, Aubrey. *Prehistoric Avebury* Yale University Press. 2002 p 140

diameters also in multiples of four as 16, 24, 32, 40, 52.1 and 60. (There are indications at Woodhenge that Burl's 52.1 was indeed intended to be the round 52 "Beaker yards"). He claimed this indicated that the counting base of the designers was four, whereas Cunnington maintained they had used a decimal count.

Maud Cunnington's decimal was correct, because Burl's Beaker yard at 73cm is 2½ times Maud's Roman foot. It is more properly a Roman step, this "Beaker yard" may be further applied to the Woodhenge geometry, inasmuch Burl's axial numbers reduction may be continued further in order to identify the overall unit of construction. All of his figures are an even number and may therefore be integrally reduced. As he stated, the axes are all multiples of four – that number should therefore be used to divide them in order to find the lowest common denominator. In this case his axes numbers become 4, 6, 8, 10, 13 and 15, these numbers, being composed of four two-and-a-half feet steps are therefore ten-feet Roman *decempedae*, this was the common module of the Roman agrimensores. "Beaker Yard" is therefore yet another extraneous term that we may throw on the scrap heap alongside the "Megalithic Yard" to avoid cluttering the landscape of metrology. "Roman step" is a universally understood module, why then, should we be introduced to it in disguise? Just as Megalithic Yard is a Belgic step. Quanta *recognition*, as practised most ably by Burl and Cunnington, should be accompanied by correct module *identification* to great advantage.

It is at this point that Alexander Thom should be taken to task in regards to Woodhenge. In his early works he sought to make comparisons of his "Megalithic yard" with acknowledged modules, one of which was the Spanish *vara*. The various Iberian varas often fit the geometry of certain of the megaliths rationally and integrally; Woodhenge is no exception. The four inner rings form the series most amenable to metrological comparisons, and if the Root Spanish vara of 2.7428571ft is divided into these axes they sum to 14, 21, 28 and 35. Each axis is a multiple of seven and seven varas are 19.2ft and this is also two ten-feet Roman perticae; thus whatever integral measure is found to fit a dimension, it will invariably lead one to the correct design module.

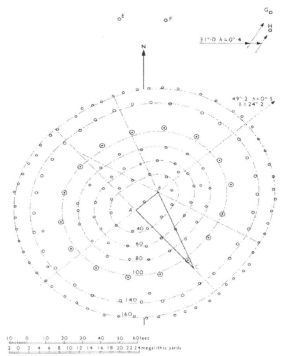

FIG. 6.16. Woodhenge, S 5/4 (51° 12′, 1° 48′). Construction superimposed: $AB = 6$, $AC = 17\frac{1}{2}$, $CB = 18\frac{1}{2}$ MY; $r =$ radii struck from $A = (P-9{\cdot}08) \div 2\pi$; $P = 40, 60, 80, 100, 140,$ and 160 MY ($P =$ perimeter).

Ring	P (MY)	r_1 (MY)	Major axis (MY)	π'	P_a (MY)
I	160	24·02	53·04	3·02	161·0
II	140	20·84	46·67	3·00	138·2
III	100	14·47	33·94	2·95	104·2
IV	80	11·29	27·58	2·90	79·9
V	60	8·10	21·21	2·83	61·3
VI	40	4·92	14·84	2·70	39·4

In this case it is the Roman pertica,[+] whence comes our term perch that is integral with all of the axes. This means that one does not merely have to find an integral measure, it must yield a rational number. The megalithic yard utterly fails to do this on any level. Thom's application of his Megalithic yard at Woodhenge is a metrological muddle.

4 The decempeda (Gr. *akaina*) is often referred to as the less specific pertica or rod, the same term is also applied to the twelve feet module.

It has just been demonstrated how the modules fit together as a clockwork mechanism, modules that may be precisely identified and classified. But his Megalithic yard varies from circle to circle without any kind of preordained regulation. Thom stated that the *"Megalithic yard which best fits Woodhenge turns out to be about 2.718, a value so close to 2.72 (used in drawing the rings) as to show that we can be quite certain we are using the identical geometric construction to that used by the builders."*[5] Certain of the geometric construction perhaps, but it is equally certain that his module is quite wrong. Thom is most clearly an accomplished surveyor and in the case of his geometrical sleuthing of the layouts of stone circles he has revolutionised archaeological attitudes and techniques regarding geometry and astronomy. But in his deduction of design modules he is totally misguided.

The diagram, left, is how Thom envisaged the basis of the geometry of the Megalithic engineers. Its departure point is a 12, 35, 37 Pythagorean triangle; the base side is upon the major axis between points he has marked A and B and is six "Megalithic yards," the adjacent A to C is 17.5 "Megalithic yards," and the hypotenuse B to C is 18.5 "Megalithic yards." His six MY, side of the triple that is upon the axis, is the separation between the regular arc centres marked A and B; to the southwest the arcs are hemispheres, to the northeast they are less than hemispheres. He has deduced the perimeters of the ellipses as 40, 60, 80, 100, 140 and 160 Megalithic yards, but one must call into question the methods that he used in order to arrive at these neat integers. The table reproduced above is taken from his *Megalithic Sites in Britain, Circles and Rings* p 75, table 6.5.

The first observation is that all of the axes are irrational; only the perimeters are given in whole numbers and in order to reach this conclusion he has listed the values that must be used for the pi ratio in the fifth column. The ratio varies with each ellipse and ranges from 3.02 to 2.7 on the distorted circles. One does not have to be a trained statistician to see the flaws in his reasoning of a consistent unit.

However, Maud Cunnington's close reckoning of increments of a Roman foot of .96 is almost childishly simple to substantiate, furthermore,

5 Thom, A. *Megalithic Sites in Britain*. Oxford at the Clarendon Press. 1971 p76

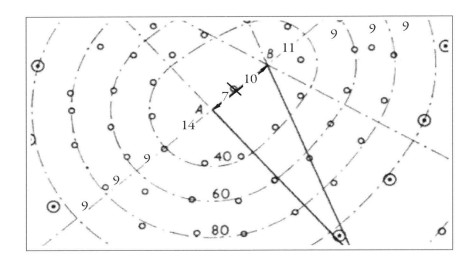

*The centre section of Alexander Thom's diagram, the superimposed numbers
show how the long axis of Woodhenge is laid out in integer increments of the
Roman foot of .96ft*

it is exactly where one would expect to find it – in the regular divisions of
the basic ground plan, the radii and spacing of the partial circular sections.

Thom's 6 MY between his centres, marked A and B, of 2.72ft is exactly
17 Roman feet of .96ft. Seventeen, being a prime, is a particularly unpleasant
integer for metrological analysis. However, it is clear from the ground plan
that the centre posthole of the whole monument lays upon the base of
Thom's triangle and would divide this base into two unequal parts of seven
Roman feet in the southwest and ten feet, the decempeda, to the northeast.
The overall length of the major axis is 150 Roman feet of .96ft, (this is also
144 English feet.) It is therefore 15 decempedae, but the spacing between
the arcs is of nine Roman feet. From the centre pole to the northeast Thom's
point B is ten Roman feet, Point B is where the north eastern arcs are struck.
They ascend to the outer ring as radius eleven and thereafter increase as 9,
9, 9, 18, and 9. To the southwest point A is seven feet from the centre post
and the first arc is radius 14 and thereafter ascends to the outer ring also
increasing by 9, 9, 9, 18 and 9 Roman feet.

There is much to be inferred from this interpretation of the Woodhenge
metrology concerning its international nature. Believed to have been

founded ca. 2,500 BC or earlier, it is contemporary with the Old Kingdom of Egypt, this is mentioned because the measure of nine Roman feet is also five royal Egyptian cubits that was the favoured foundation module of Egyptian architects. According to the Second Century Greek, Clemens of Alexandria, the Fourth Century BC philosopher Democritus claimed to have spent five years in Egypt with the architect class that he referred to as Harpedonaptai; this translates as rope-stretchers. It is widely recorded in both text and illustration that cords used in foundation ceremonies were knotted at five cubit intervals. They had thirteen knots and twelve five-cubit spaces in order to lay out the basic 3, 4, 5 Pythagorean triple, the cord is therefore a length of 60 royal cubits that is identical to 108 Roman feet.

It is obvious that the Roman Agrimensores continued the traditions of the Harpedonaptai, it is entirely probable that the Ancient Britons had used the identical techniques and it is doubtless that they used the same modules of measurement. One may note the identical practise using the same modules as the Romans and Egyptians among the Iberians by the *cordeleros* or rope stretchers. The cordel was a length of 50 Spanish varas and this was exactly 80 royal cubits and 144 Roman feet, a length perfectly preserved in the intervals of the reservoirs on the Roman aqueduct to the merida of Castile.[6]

One may go further abroad for confirmation of these facts; similar surveying methods are recorded throughout the Middle East and India but exact modules used in conjunction with the basic Pythagorean triple are recorded most reliably in China. They occur in the monument in Beijing known as The Altar of Heaven; a three-tiered structure designed to embody in all of its parts the number nine. According to still extant designs, the original modules are all reliant on this number that represents infinity and extremity.

The highest heaven was referred to as the "ninth heaven" and the lands occupied by the earliest Chinese people are referred to as the "nine states." The upper of the three terraces had a diameter of 9 zhang, (a zhang was 10 chi, a chi is a span or half cubit), the median terrace was 15 zhang and the lower was 21 zhang.

---

6 Pineiro, Mariano Estaban. *Las Medidas en la epoca de Felipe II*, Institute of Science and Technology University of Valencia. 1992 (ca)

The Altar of Heaven, Beijing

There are 9 steps in each flight of steps, therefore 3 × 9 in each of the four cardinal directions. The slabs and pillars of each balustrade are all multiples of 9. At the centre of the top platform is a circular stone that was regarded as the most sacred spot in all of China. The first paving ring around it has nine stones; the second has eighteen the third twenty-seven and so on till the final ring has eighty-one.

Centre stone of The Altar, sacred centre of the Chinese Empire

The School of Architecture at Tianjin University had conducted a survey of the monument in 1998;[7] we are therefore able to place a concrete value on the length of this particular zhang. The three terraces are 23.6m, 39.3m and 54.9m; in real money this is 77.427ft, 128.937ft and 180.118ft. Each of these totals when divided by the zhang unit is 8.603ft, 8.5958ft and 8.577ft and these are clearly 5 royal cubit measurements at lesser classifications than the 5 cubits at Woodhenge which are 8.64ft.

The use of the royal Egyptian feet and cubits in China is quite rare in the later dynasties. Fortunately, a description of the methods whereby the modules of the Altar were to be selected was preserved in the imperial archives of the Qing. When it was rebuilt in 1749, by instruction of the Emperor, Quianlong, the "customary" module of the Qing was the exact length of the Persepolitan foot of 1.0666ft; and the directive was that the chi value of the Altar construction was to be a chi that had been "anciently known" and stood in the ratio of .81 of the customary measure. This is exactly .864ft and is the value found in Woodhenge – ten of these chi are the five royal cubits that are also nine Roman feet; nine being the numerical theme of the Altar; the identical nine to the increasing axes at Woodhenge.

The cultural similarities continue with the purposes for which the monuments were constructed; at the Altar of Heaven the Emperor conducted a ceremony at the midwinter solstice and both Stonehenge and Woodhenge are similarly oriented.

Thus it is clearly demonstrated that the methods of number manipulation through the medium of the identical modules was universal. It had been practiced with fluency from a very early period, at least the Neolithic, until comparatively recent times. Although we know that there were Roman envoys in Han China and Chinese envoys in Augustan Rome, there is no indication that there was any intercourse between these disparate cultures to maintain uniformity of measure; but in this fashion their learning may clearly be traced to a common canon of reference that must have been continued through their esoteric mystery schools.

---

7   Yu Zhang (Ph.D, architect and musician, Tianjin University, School of Architecture). *Resonance: Essays on the Intersection of Music and Architecture.* Edited by Mikesch Muecke and Miriam Zach, Culicidae Architectural Press, 2007

These structures, Woodhenge and the Altar of heaven, being of concentric circular construction are most amenable to module identification and this series should be continued with another monument that had a Neolithic foundation. This is The Sanctuary on Overton hill, Avebury, for this too has concentric circles as its ground plan. It proved to be quite straightforward to isolate the design unit, as with most other sites, but the whole process is well worth describing in full because it is so informative regarding the application of the tabular system in module identification.

4.10 THE SANCTUARY

As to the dimensions of the Sanctuary the first comprehensive list of the stone and wooden-post rings in ascending order that was consulted was that of Aubrey Burl as recorded in his fully revised edition of *Prehistoric Avebury*, published in 2002. However, in the earlier edition of 1979, there was no illustrated list of the rings, merely a description of the dimensions in the "notes to pages" section at the end of the book (p 259), and the values are quite different from the subsequent revision.

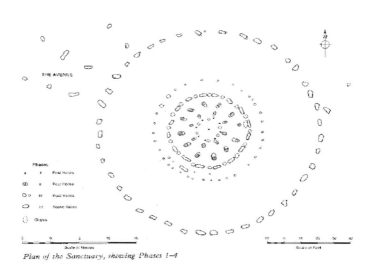

Plan of the Sanctuary, showing Phases 1–4

Ground plan of The Sanctuary showing the seven concentric circles,
two of stone and five of postholes

The list of rings that Burl gives in the 1979 edition is laid out in the order and nomenclature of Maud Cunnington's excavation of 1930. [8] He describes how *"Mrs Cunnington lettered the rings A to G, A being the outermost."* After describing the numbers and conditions of the posts he gave the dimensions as *"the timber rings D and E had diameters of 10.5 and 5.9 metres respectively."* Rings B and F were 19.8 and 4.2 metres, ring C was 14.3 metres, ring G 3.7 metres, ring A was 39.5 metres.

In the revised edition (p 134) he included a table of these rings that is quite different from the above:

Circle	No. of Holes	Size of Holes	Name of Ring	Diameter (Metres)
G	6	Big	6-Foot Ring	3.96
F	8	Small	7-Foot Ring	4.57
E	8	Big	10-Foot Ring	6.40
D	12	Big	Bank Holiday Ring	10.52
C	16	Big	Stone & Post Ring	13.72
B	32	Small	Fence Ring	19.81
A	42	Big	Outer Stone Ring	39.62

Circle	Burl 1979 edition	Burl 2002 edition	Wainwright with Longworth [9] 1971
G	12.139ft	12.992ft	13.5ft
F	13.78ft	14.993ft	14.17ft
E	19.357ft	20.997ft	19.83ft
D	34.45ft	34.514ft	32ft
C	46.916ft	45.013ft	46.5ft
B	64.96ft	65ft	64.5ft
A	129.6ft	129.987ft	129ft

For a multitude of reasons metrological analysis is far more straightforward if dimensions are expressed in English feet and is slightly more accurate if the English conversion from the metric (as opposed to the

---

8 "The Sanctuary on Overton Hill, Near Avebury" in the *Wiltshire Archaeological and Natural History Magazine* vol. xiv

9 Measurements quoted by Chamberlain, A. T. "Units of Measurement in Late Neolithic southern Britain", from the book: *From Stonehenge to the Baltic. Living with Cultural Diversity in the Third Millennium BC*, Archaeopress 2007. (Oxford, *British Archaeological Reports International Series* 1692)

rounded American) is employed, the equivalence being 3.28084275 feet to the metre. Burl's measures in English are then the first two columns as above.

It is not made clear in either edition of Prehistoric Avebury exactly how Burl came by his given dimensions, it would appear from the text of 1979 that he used the figures of Maud, whose nephew, Col. Robert Cunnington, a competent military surveyor, made the measurements. In the literature upon the subject, certain archaeologists have alleged crudeness on the part of both Maud and Robert regarding their methods, but the present author would as well trust a skilled surveyor with a piece of string as an archaeologist with a theodolite.

The discrepancies between the two editions are not huge, but in some cases would certainly be sufficient to alter the derived module. When dealing with postholes there is always a good margin of play, not that this precludes accurate module identification once a design theme is recognized; and that is what is sought.

It was the later, 2002 edition, that was first consulted and it was immediately noticed how most often the listed diameters were closely integral with the English foot and that circles G and A would be related as one to ten. Six of the circles having near English foot integers as A 130, B 65, C 45, E 21, F 15, and G13, most would regard as – if not conclusive as to the design module – then at least probable.

The problem with accepting this is that the numbers are unsuitable to circular designs. What is meant by this, is that one will invariably find a module by a division of the diameter of a circle by 14; the fact that the least and greatest diameters of The Sanctuary are given as 13 and 130 feet and if they are divided by 14 the resultant module is .92857ft, in other words 13 to 14 of the English/Greek foot. This would appear to conform to the rules of metrology by virtue of the fact that it is a unit fraction connection. However, this solution may be instantly dismissed on the grounds that 13 is a prime number and cannot therefore be part of an integrated system as a module.

One therefore seeks a length closest to the derived value that is a proven module. In this case the closest to .92857ft would be the Standard Geographic Iberian pie or one third of the vara of 2.7806ft at .926866ft. It is a truly ancient measurement that predates Roman occupation of Iberia. (This is a well provenanced value found in many Spanish and Portuguese colonies, recorded from Peru, Mexico, Buenos Aires, Manila, Curacao, and Brazil. But most convincingly defined in the Texas land grants as 33.4 inches that they rounded to be exactly 33 1/3 inches for the sake of simplifying conversions to the more familiar acres from square leguas; the precise Spanish length would be 33.367 English inches which is .84752m.)[10] This is the module that appears to be the most convincing fit of The Sanctuary circles at:

	Pie	Feet	Metres
G	14	12.9 76	3.96
F	15	13.903	4.24
E	21	19.464	5.93
D	35	32.44	9.89
C	50	46.343	14.1
B	70	64.88	19.77
A	140	129.761	39.55

The first column is the circle, the second column is Iberian feet, the third English feet and the fourth metres in terms of the round numbers of the Iberian pie of .926866ft or 28.25cm.

In order for the circle D to be the round number of 35 pie the figure of 32 feet as tendered by Wainwright and Longworth has been used. This lesser value than those forwarded by Burl is legitimate because this circle is of larger double postholes that occur in two of the rings; the nature of these posthole circles is most adequately and amusingly dealt with by Mike Pitts.[11]

That the circles seem to be unified by a common geometrical theme uniting the timber and the lithic elements would imply that developments over a very long time had adhered to and had elaborated upon the original timber design. The fact that Mortlake, Fengate and Grooved Ware sherds were found in the lower fills, and the later Beaker Ware from rings D and E implies that this development was spread from at least the Neolithic well

10 Doursther, Horace: *Dictionnaire universel des poids et mesures anciens et modernes* Bruxelles, M. Hayez, imprimeur de l'Académie royale, 1840.

11 Pitts, M. "Return to the Sanctuary", *British Archaeology* 15 (2000), 15–19.
 Pitts, M. *Hengeworld*, Random House, 2011

into the Beaker phase, probably over a period of 600 years.[12] It would also seem that some of the circles—both stone and post—were contemporary from the geometry suggested by the rings of stone – A, and posts – B, as illustrated in the following diagram.

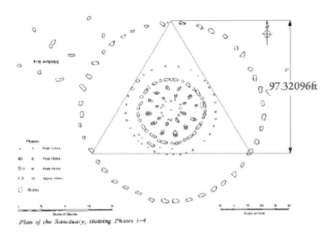

Plan of the Sanctuary, showing Phases 1–4

The outer stone circle governing the outer posthole geometry.

The vara of three of these Sanctuary Iberian feet is exactly 35 to the inner sarsen diameter of Stonehenge and 110 to the perimeter. Although it is compatible with the Stonehenge dimensions it was not the module of design; much as the vara of Castile (shorter by a factor of 1.01376) was shown to be compatible with the Woodhenge measurements. The Iberian foot is quite rare in Megalithic circle design, one other instance has been noted in one of the Lake District complex of circles, Burnmoor D as Thom listed it, whose constituent foot is the 440th part less than the Sanctuary module. But the method outlined in this Sanctuary description is undoubtedly the correct way to approach module identification. Battalions of statisticians cannot possibly identify Megalithic quanta, for the simple reason that they seek a single unifying unit, which does not exist in ancient metrology.

The fact that this Iberian foot is the module of The Sanctuary design is borne out most convincingly by a final resolution to the geometry. This

---

12   Thomas, Julian. *Understanding the Neolithic*, Routledge, 1999

concerns the perimeter of the outer stone circle; the diameter of this circle is 140 Iberian pie and this is 129.76128 statute feet. The perimeter of this circle by using 22/7 is 407.821+ft; there are 42 stones in this outer circle and this total divided by 42 equals 9.71+ft. This is exactly the decempeda of ten Root Geographic Roman feet. These facts underscore that we are dealing with a sophisticated numerical metrological scheme that is exactly expressed in a vast numbers game that was practised by the Megalithic engineers. A very similar resolution in the outer perimeter of the Mount Pleasant timber rings will shortly be demonstrated.

The diagram opposite showing the integration of the outer stone and outer timber circles of the Sanctuary through the inscribed triangle, has a further function of illustrating, with exactitude, the fact that they were using a unified system throughout the Megalithic arena. A triangle inscribed in a circle accurately divides the diameter of the circle into four, this is three parts as the height of the triangle – to one part as the distance between the base and the perimeter. It is the height of the triangle as three quarters of the diameter that proves to be highly significant as a module. It is a plethron of 100 Roman feet, and the Roman foot in question is exactly the Standard Geographic value i.e. .96ft (Root) x 1.10376 = .9732096 (Standard Geographic.)

This gives a perfect insight into the nature of the integrated system as a whole, both in precise modules and the methods of application, because this plethron is exactly the inner diameter of the sarsens of Stonehenge as identified by Petrie – to within one twentieth of an inch. Further to this, the fact that the separation between the outer stones of the Sanctuary is the Root Geographic Roman decempeda of 9.71+ft; it is this plus its 440th part that is the foot of the diameter value. This is merely what happens to number quite spontaneously and it would be naive to believe the ancient geometers were unaware of this.

These facts underscore that one must never come to the data with ideas of applying a preconceived unit. One deduces the modules of circles by applying the ratios that we know have been used, rational numbers result, and these are invariably found through a division of the diameter by seven or multiples thereof. Although Aubrey Burl had identified a quite accurate value for the Roman step as his "Beaker yard;" his negative attitude to prehistoric mensuration is betrayed by his slack groping for a module of The Sanctuary.

He suggested that Ring C, which he believed was 13.72m in diameter, may have been composed of eight "body-fathoms", as he termed it, of 1.74m. This would be 13.92m and has nothing to do with The Sanctuary modules, although, once again, he has closely hit upon a proven value that is directly related to his Beaker yard. It would be a fathom of six Roman feet, just as his "yard" was a step of 2½ Roman feet. This fathom at the Root Roman of Woodhenge would be 1.756m that differs from Burl's value by a factor of 1.008, which would make Burl's fathom 1.7417m. He is therefore accurate to a known variant within .1 of a percentage point.[13] At its basis of the foot length, it is obviously a module of the Sanctuary as we have clearly demonstrated, but *nothing at all* as Burl describes it.

In spite of the fact that Aubrey Burl is justifiably regarded as a great authority on the Neolithic-Bronze Age societies of Europe, he refuses to believe what all of the evidence from those ages screams aloud; that of a continent wide, highly regulated intellectual organisation. Whereby enormous feats of landscape engineering and megalith building were carried out to strict specifications of uniformity. This would obviously have necessitated powerful central authority, both religious and political that liaised with others of a similar nature over vast geographic distances. His implied disrespect of these ancient and highly accomplished ancestors is betrayed by his baseless and insufferably patronising remarks that he lets slip in his *"Prehistoric Avebury."*

Firstly, in paraphrasing a statement by the 17th Century English philosopher, Thomas Hobbes, that before civilization life was *"solitary, poor, brutish and short"* implying that this was the condition of men in the Neolithic (although he did lighten up on the "brutish.") Then in his description of the metrology of The Sanctuary he came up with the following pearl – *"Similarly ring D, 10.52m, may originally have been made up of twelve body-yards of the same unit. (half of 1.74m) But none of the other rings G, F, E, B, A conforms to this "Yard" implying several stages of construction, something quite consistent with the work of subnumerate peasant craftsmen living in prehistoric Britain who were not much concerned*

---

13 This lesser Roman fathom is also related to the Root royal Egyptian cubit through the canonical man artist grid. Every man is six feet tall and the Egyptian canonical man is a stature of 3 ⅓ royal cubits therefore: $12/_7 \times 1/_3$ = 5.71428ft this divided by 6 = .95238ft and this is accurately the lesser Root Roman foot or one sixth of Burl's "body fathom."

with nice measurements as long as their buildings were solidly constructed."
Fortunately such absurd ideas as subnumerate are attitudes that are dying
out along with the archaeologists of Burl's generation. This brashness is
best regarded as the establishment reaction to the wilder of the theorists
that had been ushered in by Thom, who made such a clamour in the
nineteen sixties and seventies. The maturing archaeologists who are now
on stream have grown up with the background noise of that New Age
sensationalism and can more easily winnow the wheat from the chaff; they
are becoming more receptive to the well reasoned arguments regarding
ancient mathematical sophistication.

Two such modern archaeologists are the previously mentioned
Andrew Chamberlain and Michael Parker Pearson who coauthored
"Units of Measurement in Late Neolithic Britain." They at least give
credence to a multi unit application, although they show reticence at
positive identification of the modules that they uncover, they are also
guilty of trying to fit these preconceived units onto other monuments
under consideration. A more efficient way of module identification, as has
been demonstrated, is seek the module that is self evident in the individual
structures through rational numbers that emerge from known ratios; such
as the approximations of $\sqrt{2}$, pi and phi and so forth, and these should
then be applied to the measured lengths expressed in English feet.

Refreshingly, they express dimensions in terms of this statute foot
as well as the metric equivalent, they at least realise it is easier to see the
connectedness of the modules when articulated in this way.

They use the Roman foot of .96ft but often term it the "short" foot,
and they use the "long" foot of 1.056ft that is universally recognised as
the Persian foot. They also make reference to an unspecified module that
is eleven to ten of the statute but fail to term it the Saxon foot. Specific
terminology is necessary in such matters; even though these received
nationalistic terminologies may be deficient, it is all we have at present
and there would at least be rough common agreement on what is the foot
module under consideration if it is identified.

Although they go much further than others in the field regarding
Neolithic metrology it remains lip service, and as do all others who attempt
to understand the matter, they defer the problem stating that the *"subject
warrants further investigation."* Kicking the can down the road, so to speak.

4.11 MOUNT PLEASANT, DORCHESTER, UK

After that little rant we may continue with another highly significant post ring construction, Mount Pleasant, Dorchester. Although of a very different scale to The Sanctuary, it has many similarities, Pollard stated of it:

> *"In terms of its architecture and organization of space, the Sanctuary finds a close parallel with contemporary timber and stone settings within Site IV, Mount Pleasant (Wainwright 1979). Both structures incorporate cruciform corridors, in the case of Site IV aligned upon a cardinal axis, dividing the settings into four equal quadrants with a central circular open area. The internal symmetry of both structures sets them apart from contemporary timber circles such as the Southern and Northern Circles at Durrington Walls, Woodhenge and Balfarg."*[14]

The table below is taken from Pollard after Wainwright. The rings of this monument are quite irregular meaning that they do not adhere to perfect circles; but they do conform to a perfectly circular scheme inasmuch they are all struck from a common centre.

If they are regarded from the point of view of the individual quadrants, as shown in one of Pollard's tables, the data may appear more irregular that it actually is. The table below, that shows the north-south and east-west diameters is quite sufficient to display a certain regularity in the plan and one may deduce the module from the scheme. It is obvious from the numbers that emerge, even when they are expressed in metres, that the designer was increasing the circles by six units.

The nearest whole number beginning with the inner circle E is 12m, D is 18, C is 24 and B is 30; an obvious progression by modules close to six metres. This regularity breaks down with the final circle, A, being 38m, and this ought to be 36 metres in order to complete the regular progression. As this amounts to a full two metres difference on the diameter, it is too much to be accounted for as error by the designer – or the reporter.

---

14 Wainwright, G J. *"Mount Pleasant, Dorset: Excavations 1970-1971. Incorporating an account of excavations undertaken at Woodhenge in 1970."* Society of Antiquaries of London. Thames and Hudson 1979

Pollard, Joshua. *The Sanctuary, Overton Hill, Wiltshire: A Re-examination.* Proceedings of the Prehistoric Society, 1992

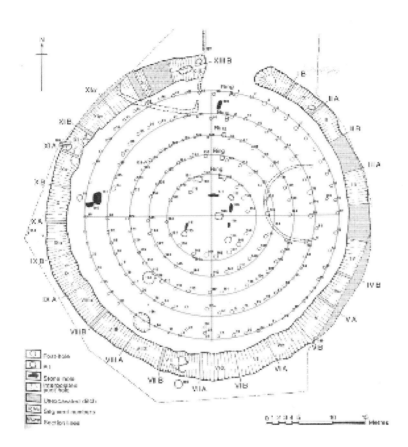

Mount Pleasant timber circles showing the common centre

	Mean diameter		No. of posts in each ring						
Ring	North to South	West to East	Excavated	Presumed	Replacements etc.	Average spacing	Average post-hole diam. (cm.)	Average post-hole depth (cm.)	Average post diam. (cm.)
A	37·30	38·00	45	52	3	2·10	43	26	28
B	29·50	30·80	46	48	—	1·93	45	30	—
C	24·00	24·60	35	36	2	1·98	44	29	—
D	18·30	18·20	24	24	—	2·21	48	34	—
E	12·20	12·50	16	16	2	2·11	53	36	—

In spite of the obvious spatial irregularities, there is a remorselessly logical play in the numbers that emerge when the module that the numbers are expressed in is identified. Mount Pleasant is a perfect example of the axiom that one comes to the simple geometry with no preconceived thought of trying to fit a module onto it, one then deduces the module from the ratio(s) in combination with the measured length.

It is the English foot that is the most useful tool of comparison to identify the emerging module, as it is directly related to the majority of ancient foot-lengths; and indirectly related to all the rest. One then converts from the metric scale at 3.28084275 feet to the metre. (This is the Benoit and Chaney definition of 1898.) Starting with the inner ring the given metric lengths – as described by Wainwright – when converted to feet are as follows:

AVERAGES OF NS AND EW DIAMETERS

1st, inner ring	12.35m	40.518ft
2nd ring	18.25m	59.875ft
3rd ring	24.3m	79.72ft
4th ring	30.15m	98.917ft
5th outer ring	37.65m	123.524ft

The regular progression of the first four circles as 12 - 18 - 24 - 30 in the metric, is then 40 - 60 - 80 - 100 as English feet. In order to identify the primary module of design one simply divides the feet diameters by fourteen, the smallest centre circle would then be a measure of near 3 Roman feet of .964ft, and the 4th inner ring, the longest in the "series" gives a little over 7 feet as its fourteenth part, this is around 7.5 ft of .94ft. Immediately, one has reduced the hunt for quanta to a single choice – the design module is either a Roman foot or a Samian foot.

Further divisions of the diameters reveal that the most likely candidate for the design module is the Root Samian foot of .9428571ft. It works as so:

	Samian ft	English feet
1st, inner ring	42	39.6
2nd ring	63	59.4
3rd ring	84	79.2
4th ring	105	99
5th outer ring	132	124.457

As one fourteenth part of the increasing diameters they maintain a known module, the first is three Samian feet, or yard, the next is 4½ feet and this is three cubits called a *xylon* by the Greeks, it is used in the nine feet multiples. The next is the six feet fathom and the fourth diameter is fourteen five-cubit modules. Seven and a half feet – or five cubits, has many applications as a module, it is most commonly encountered in the diameters of 100 feet or "plethron" circles.

Although the final, outer circle, is not part of the ascending series, at 132 Samian feet it maintains integrity, but if it were a continuation of the series it would be 126, in the following fashion. The first ring at 42 is 2 to 3 of the second; the second is 3 to 4 of the third; the third is 4 to 5 of the fourth; the fourth should therefore be 5 to 6 of the fifth at 126; it is not, at 132 it is 6 feet longer.

This 6 feet length overall is one third of pi at 22 to 21 of the 132 to 126. This makes the perimeter of the outer circle 391.0217ft. If this circle is divided by 360 it is 1.08617ft and this is the identical Belgic foot as found in the plethron of the Stonehenge station stone quadrilateral.

The Belgic modules are commonly found in Neolithic remains and Brochs, they persisted in Britain throughout the Middle Ages. The Samian foot also had worldwide distribution – it was the official foot module of many German cities; it is the 800th part of the base of the Great Pyramid (Herodotus), its two feet cubit was the *pyk belady* that has only been abandoned in Egypt since metrication. It is accurately found in China, and in Japan its cubit is the arm of the *sashigane* or carpenter's square.

4.12 DURRINGTON WALLS, SOUTH CIRCLE

This regularity of design in the quartet of timber rings that was written up by G J Wainwright and commented on by Andrew Chamberlain and Mike Parker Pearson, proved more difficult to find with the final, Durrington Walls, South Circle.

As with the other timber rings there is a certain ambiguity in the reported measurements, as is to be expected with such large reference points as substantial postholes. Although with the previously described rings there were very obvious patterns that emerged that enabled precise identification of the design modules. A consistent pattern proved more elusive in Durrington Walls; this would be borne out by Wainwright's illustrations of the four timber circles. There are discernible regularities in Woodhenge, The Sanctuary and Mount Pleasant – that are lacking in the more haphazard Durrington Walls.

Just as was noted with the divergent values that Burl gave for the Sanctuary dimensions in different publications of his work on Avebury, the same observation may be applied to Wainwright, inasmuch that in his 1971 work – *Durrington Walls* – his dimensions for Durrington differ from the dimensions that he included in his 1979 *Mount Pleasant, Dorset* work. From both works the six diameters of Durrington are reported as follows:

1971	1979
10.73m	10.7m
15.09m	15.2m
22.55m	22.9m
29.01m	29.3m
35.46m	35.7m
38.46m	38.9m

As can be seen from these comparisons, the differences are not great, up to around half a metre. This is to be expected had the reports been submitted by different authors but in the case of such esteemed scholars as Burl, and Wainwright in this case, it betrays carelessness regarding the importance that they assign to metrology. This is because in other respects they are so meticulous in describing all else, from potsherds, depths of postholes, fragments of bone and all of the other minutiae down to grains of pollen that are encountered in excavations – they are anything but so cavalier.

Phase	Structural Component	Diameter (metres)	Diameter (stat. ft)	Diameter ('long' ft)	Modular Diameter	Absolute Error
2	2F	10.73	35.20	33.33	?	-
2	2E	15.09	49.52	46.90	?	-
2	2D	22.55	73.99	70.06	70	0.06
2	2C	29.01	95.17	90.12	90	0.12
2	2B	35.46	116.35	110.18	110	0.18
2	2A	38.46	126.19	119.50	120	0.50

Above, is the table composed by Chamberlain and Pearson in their article *Units of measurement in Late Neolithic southern Britain*, it is based upon the Wainwright with Longworth values of 1971. They use quite good metric conversion values to arrive at their statute feet (unlike the majority of archaeologists).

Their conclusions are obviously based upon experimentation with the values that they have deduced elsewhere and they have produced a convincing series of integers for the rings based upon the "long foot" (which is the Root Canonical Persian foot of 1.056ft, assuming a Root of 1.05ft). How the present author approaches the solution is quite different. The problem would be set out as follows.

	d stat. feet	1/14th d	perimeter
a	35.1	2.507	110.33
b	49.86	3.562	156.73
c	75.131	5.366	236.126
d	96.128	6.866	302.118
e	117.126	8.366	368.11
f	127.166	9.12	401.1

Only two significant numbers emerge from this exercise, the first is the 14th of the diameter of circle *a* is the English/Greek step of 2½ft, the other is the diameter of circle *d*, it is one hundred Roman feet. On experimentation, the other diameters had shown no relationship to the Roman foot but are very closely integral with the English step at 14, 20, 30, 38, 46 and 50. The margins of error are identical to the Persian foot solutions tendered in the Chamberlain – Pearson solutions. As all of the circles are closely integral with the English foot and step, then this should be the answer.

Because of this, it was believed that this set of circles had no elegant resolution because rings *d* and *e* at 38 and 46 steps were double prime numbers and these are notoriously difficult to fit into cogent numerical schemes; it was looking as though these particular timber rings may have been roughly paced, as the detractors of megalithic measures had always claimed that the circles were.

On experimentation with perimeters in order to find more convincing integers, at first, the ratio of 22/7 was applied to these derived diameters resulting in the following series as the perimeters in English feet. Only the first, the 35 feet diameter yielded a recognisable module, this was 100 Saxon feet, a plethron. Circle *a* 110ft – *b* 157.142ft – *c* 235.714ft – *d* 298.571ft– *e* 361.428ft – *f* 392.857ft. The Saxon foot was not replicated in any of the other perimeters, but they were all closely integral with divisions by the common Egyptian foot. But in order for this to be exact, all of the perimeter numbers had to be reduced by the 3025th part.

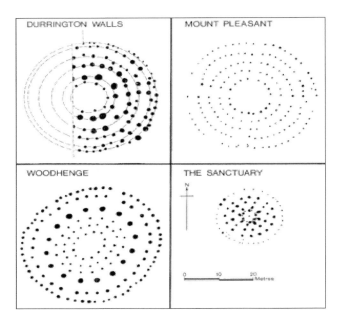

Comparative plans of the large buildings at Durrington Walls,
Mount Pleasant, Woodhenge and The Sanctuary (Wainwright p 226)

What this effectively means, is that the pi value must be altered from 22/7 to the more accurate 864/275, in order to give the correct values. For example, the Saxon foot is modified Sumerian measurement; therefore the Saxon 1.1ft reduced by the 3025th part is Standard Sumerian of 1.09963ft. 100 Sumerian feet is also is also 112 Standard common Egyptian feet of .981818ft, and this proves to be the rational integer module of the Durrington Walls timber circle perimeters. It is the most commonly acknowledged value of the common Egyptian foot by virtue of the fact that it is 16 digits of the 28 digits of the royal cubit of 1.71818ft that had been identified by Petrie as the design module of the Great Pyramid. The perimeters would be as so in terms of these feet:

diameters English feet		steps	perimeters English feet	perimeters common Egyptian feet
a	35	14	109.9636	112
b	50	20	157.0909	160
c	75	30	235.6363	240
d	95	38	298.4727	304
e	115	46	361.309	368
f	125	50	392.7272	400

As with the other timber circles, the final, outer perimeter, is a very rational 400, in this case – common Egyptian feet; in the case of the Sanctuary it was 140 Iberian feet and at Mount Pleasant it was 360 Belgic feet. It would be fair to assume that the designers of all of these monuments were engaged in the same sacred numbers game – that is virtually undetectable when expressed in the decimal notation.

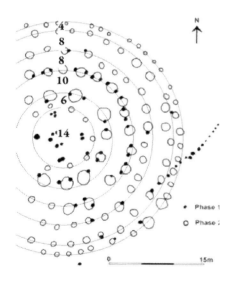

Diagram of Durrington Walls South circle, showing the increase in diameters in terms of the English/Greek step[15]

15 Gibson, Alex. *Stonehenge and the Timber Circles of Britain and Europe*, Tempus 2005

4.13 Conclusions

The common Egyptian foot is quite often found in megalithic circle design, at least a dozen instances are listed whose variants may be most accurately identified. The value at Durrington Walls is the Standard value of .98<u>18</u>ft and to compare it with the value found at Stonehenge it must be twice increased by its 175th part to be .993071ft. In terms of its relationship to the English foot it is 48/49 increased by 441/440 – Durrington; and 48/49 x 1.01376 – Stonehenge.

In lesser monuments such as these timber circles, regardless of the fact that they may have evolved over centuries (the Sanctuary), a single design unit is precisely identifiable; this implies continuity of culture. It may be that differences in pottery styles does not mean the intrusion of an alien people, it may simply be indicative of an evolution in manners or fashion. If this were not the case, and foreigners *had* displaced the natives, then both must have inherited a common culture of mathematics and ritual in order for developments on a unified theme to process so seamlessly.

In the more important monuments, their complexity precludes the identification of a single design module. Regarding Stonehenge, Chamberlain and Pearson remarked that – *"We suspect that a single unit of measurement was used for any given phase of construction, a proposition that is testable at sites like Stonehenge where the chronology of construction is now reasonably well established."*

This is clearly not the case, firstly because all of the modules in all of the nations, were used concurrently; and then all of the Stonehenge developments were continuations on a singular vision that necessitated many interlinked modules regardless of the time of construction. For instance there are three perfect examples of the *plethron* that are preserved in the stones, the inner diameter of the sarsens is 100 Roman feet, its outer diameter is 100 common Greek feet and the shortest sides of the station stone quadrilateral are 100 Belgic feet. The fact is, that the circular elements of Stonehenge are not linked together by a common module; but the unification is by a common *classification* term, and this includes the many modules in the monument that are additional to these plethra.

In the case of Stonehenge the underlying theme is the classification term *Standard Geographic* it is the formula 1.01376. The Root value of the Roman foot is .96ft, the Root value of the common Greek foot is 1.028571ft and the Root value of the Belgic foot is 1.0714285ft. These are respectively 24 to 25 of the English/Greek foot, 36 to 35 and 15 to 14. When multiplied by 1.01376 these root numbers achieve the values of the Stonehenge proportion, .9732096 Roman, 1.04272ft common Greek and 1.08617ft, Belgic. This is also the classification formula of the Sanctuary, in this case it is the Root Iberian foot of .9142857ft multiplied by the same 1.01376 to equal .926866ft.

This is also the same classification as the modules of the Parthenon and this temple also exemplifies *plethra* in a very obvious way, because this temple is known as *Hecatompedon* which means "hundred footer." The lower step of the stylobate is one hundred Sumerian feet, the second step is one hundred Persian feet and the top step is one hundred Greek feet at 101.376ft, (Rottlander) the length of the *naos* is one hundred Roman feet; the naos, at 97.32096ft is identical to the inner sarsen diameter at Stonehenge and the triangle of the Sanctuary geometry.

These modules that are duplicated in each building, may also be assumed to have served an identical purpose in each building and this would be the permanent preservation of the standards of measurement, against which any merchant, tradesman, priest or scientist could test his own instruments. Testing of instruments may even have been a feature of the ritual gatherings or gorseddau. With an equal possibility that various tribal or clan groups within a nation preserved their own allotted standard from the range of measures. These comparisons may have been how the modules survived unchanged through the millennia. But that is enough speculation.

Such insights as the all important *module recognition* are made possible by dint of practise and research, an often onerous task that new students of the study do not now have to undergo. This is because the foot and potential foot modules have been listed and classified. Any module that does arise in architectural comparisons may therefore be easily checked for validity and precise identity.

March 2016

4.14 APPENDIX OF THE MEGALITHIC RING DIMENSIONS

This list is Thom's circle diameters taken from his surveys in the book "Megalithic Rings" that he co-authored with Aubrey Burl. The first column is the diameters in feet, the second is the diameter divided by 14, the third is the identification of the module, the fourth is the explanation and identification of the constituent foot, the fifth is the classification of the foot (RG =Root Geographic, SG = Standard Geographic etc, see page 5), the final column is the difference overall. Note the five cubit module; this was commonly used in long measurement by the Egyptians. Also note the steps and paces, the Megalithic yard and fathom should appear in these but is notably absent. Largely because it doesn't exist.

Megalithic Rings

		Diameter ft	D/14	Module	Analysis		Identification	Overall Error from Thom. inches
Nine Ladies		35.5	2.537	step	2½ ft	of 1.01376ft	SG Greek ft	½
Elva Plain		113	8.0714	5 cubit	7.5ft	of 1.07755 1ft	RC Belgic ft lesser	1½
Sunken Kirk		92	6.571	fathom	6ft	of 1.09486ft	SG Belgic ft greater	½
Burnmoor A		70	5ft	pace	5ft	of 1ft	Rt English/Greek	0
Burnmoor B		49.7	3.55	yard	3ft	of 1.1827ft	RG Russian ft	3
Burnmoor C		54.6	3.9	double cubit	4ft	of .97402ft	RR common Egyptian	½
Burnmoor D		52	3.7142	double cubit	4ft	of .92686ft	SG Iberian ft	1
Blakely Moss		54.8	3.914	double cubit	4ft	of .97959ft	Rt common Egyptian	½
Tarnmoor (con)	inner	89	6.3571	fathom	6ft	of 1.0584ft	SC Persian	1
	outer	96	6.8571	fathom	6ft	of 1.14285ft	Rt Royal Egyptian	0
Gunnerkeld (con)	inner	50	3.5714	?				
	outer	102	7.2857	5 cubit	7.5ft	of .971ft	RG Roman	0
Oddendale (con)	inner	24	1.71428	cubit	1.5ft	of 1.142857ft	Rt Royal Egyptian	0
	outer	86	6.14285	fathom	6ft	of 1.02272ft	RR common Greek	1
Birkrig (con)	inner	27.7	1.97857	cubit	2ft	of .98742ft	SC common Egyptian	½
	outer	83.3	5.95	fathom	6ft	of .99307ft	SG common Egyptian	0
The Hurlers	center	136.6	9.761	decempeda	10ft	of .9752ft	SG+Roman	1
	north	113.6	8.1185	5 cubit	7.5ft	of 1.08245	Std Belgic lesser	0
	nouth	104.6	7.476	5 cubit	7.5ft	of .99353ft	SG+common Egyptian	1

Name							
Nine Stones	49.6	3.5428	yard	3ft	of 1.18ft	SG Russian	½
Stripple Stones	147	10.5	decempeda	10ft	of 1.05ft	Rt Persian	0
Treswigger	108.3	7.7357	5 cubit	7.5ft	of 1.0309ft	Std common Greek	½
Leaze	81.5	5.82	fathom	6ft	of .971ft	RG Roman alt	½
Nine Maidens Cambourne	53.6	3.828	double cubit	4ft	of .95782ft	Std Roman lesser	½
Nine Maidens Ding Dong	71.6	51.2	pace	5ft	of 1.02272ft	RR common Greek	0
Merry Maidens	77.8	5.557	pace	5ft	of 1.1111ft	Rt archaic English	½
Trezilbbet	106.2	7.5857	5 cubit	7.5ft	of 1.01146ft	SG Greek	0
Grey Wethers	104.5	7.464	5 cubit	7.5ft	of .994318ft	RR Greek	1
Brisworthy	81.4	5.81	fathom	6ft	of .96768ft	SC Roman	1 ½
Ringmoor	41.4	2.957	yard	3ft	of .987428ft	SC common Egyptian	½
Stanton Drew	372.4	26.6	10 steps	2.5ft	of 1.06648ft	SG Persian	2
Rollright Stones	103.6	7.4	5 cubits	1.5ft	of .98742ft	SC common Egyptian	1
Temple Wood	44.5	3.178	yard	3ft	of 1.0584ft	SC Persian	½
Machrie Moor	54	3.857	double cubit	4ft	of .96548ft	RC Roman	½
Archnagorth	44.9	3.2	yard	3ft	of 1.06686ft	SG+Persian	1
Aquorthies Manor recumbent	4.5	4.5	pace	5ft	of .9ft	Rt Assyrian	0
Sheldon of Bourtie (con) inner	53	3.78	double cubit	4ft	of .945ft	Std Samian	1
outer	108.4	7.742	5 cubit	7 ½ ft	of 1.0309ft	Std common Greek	2
Westerton	49.5	3.536	yard	3ft	of 1.176ft	SC Russian	1 ½
Loanhead of Daviot (con) inner	54.4	3.8857	double cubit	4ft	of .971ft	RG Roman	½
outer	68	4.857	pace	5ft	of .971ft	RG Roman	½
Tyrebagger	59.3	4.325	double cubit	4ft	of 1.0584ft	SC Persian	½
Sunhoney recumbent	83	5.928	fathom	6ft	of .9874ft	SC common Egyptian	½
West Mains recumbent	66.8	4.7714	pace	5ft	of .95238ft	Rt Roman lesser	½
Garrol Wood recumbent	58.5	4.1785	double cubit	4ft	of 1.04403ft	RR Persian	½
Cullerlie	33.3	2.3786	step	2½ft	of .95238ft	Rt Roman lesser	0
Tarland	74.1	5.2928	pace	5ft	of 1.0584ft	SC Persian	0
Midmar Church recumbent	56.8	4.0571	double cubit	4ft	of 1.01376ft	SG English/Gk	0
Tillyfourie Hill recumbent	72	5.1428	pace	5ft	of 1.02857ft	Rt common Greek	0
Aquorthies Kingousie (rec/con)	49.7	3.55	yard	3ft	of 1.18272	SG Russian	½
outer	75.1	5.36428	pace	5ft	of 1.071428ft	Rt Belgic lesser	1 ½
Clune Wood recumbent	56	4	double cubit	4ft	of 1ft	Rt English/Gk	0

Site	Qualifier			Unit		of	Type	
Carnhousie House pair Northern		27	1.92857	cubit	2ft	of .96548ft	RC Roman	½
	Southern	84	6	fathom	6ft	of 1ft	Rt English/Gk	0
Milltown		91.7	6.55	fathom	6ft	of 1.09237ft	RG Belgic (Drusian)	½
Urquart		110	7.85714	5 cubit	7.5ft	of 1.047619ft	RC Persian lesser	0
Gask (con)	inner	82.9	5.9214	fathom	6ft	of .987428ft	SC common Egyptian	½
	Outer	119.4	8.52857	5 cubit	7½ft	of 1.13789ft	RG Nippur	1
Farr West (con) tpe A	inner	66.8	4.771	pace	5ft	of .95238ft	Rt Roman lesser	1½
	outer	113.2	8.0857	5cubit	7½ft	of 1.07755ft	RC Belgic lesser	½
River Ness (con)	inner	30.1	2.15	cubit	2ft	of 1.07386ft	Std Belgic lesser	½
	outer	69.1	4.9357	pace	5ft	of .9874ft	SC common Egyptian	½
Drumandow		89.1	6.36428	fathom	6ft	of 1.06203ft	RG Persian	1½
Loch Mannoch		21	1.5	cubit	1½ft	of 1ft	Rt English/Gk	0
Cambret Moor		82.1	5.8642	pace	5ft	of 1.17333ft	RC Russian	½
Auldgirth		100.2	7.157	5 cubits	7½ft	of .9545ft	Std Roman lesser	½
Loch Buie pair	lesser	21.75	1.5536	cubit	1½ft	of 1.03448ft	RC common Greek	½
	greater	44.1	3.15	yard	3ft	of 1.05ft	Rt Persian	0
Forse		157.5	11.25	decempeda	10ft	of 1.125ft	Rt Nippur	0
Latherton Wheel		188.3	13.45	pertica	12ft	of 1.12ft	SC Archaic English	1½
The Mound		24.5	1.75	cubit	1½ft	of .75ft	Rt Black cubit	0
Shin River (pair)	lesser	13.6	.97142	foot	1ft	of .971003	RG Roman	0
	greater	24.5	1.75	cubit	1½ft	of .75ft	Rt Black cubit	0
Brodgar		340.9	24.35	double pertica	24ft	of 1.01376	SG Greek	3½
Leys of Marlee		49.4	3.5285	double cubit	1½ft	of 1.764	SC Black cubit	0
Croft Moraig (con)	inner	41	2.928	yard	3ft	of .97542	SG+Roman foot	½
	outer	58.5	4.17857	double cubit	4ft	of 1.047245ft	SG common Greek	1½
Shianbank pair both		27.5	1.9642	cubit	2ft	of .981818ft	Std common Egyptian	0
Gors Fawr		73.2	5.228	pace	5ft	of 1.04403ft	RR Persian	1½
Castell Garw		43.8	3.128	yard	3ft	of 1.04403	RR Persian	½
Y Pigwyn		76.3	5.45	pace	5ft	of 1.09237ft	SG Belgic (Drusian)	1
Y Pigwyn (con)	inner	24.3	1.7357	cubit	1½ft	of 1.15595ft	SG royal Egyptian	½
	Outer	43.7	3.1214	yard	3ft	of 1.04036ft	RG common Greek	0
Maen Mawr (eccentric)		59.75	4.2678	double cubit	4ft	of 1.06666ft	Rt Persepolan	0
Ynys Hir		58.6	4.1857	double cubit	4ft	of 1.045094ft	SG+ common Greek	½
Grey Hill		32.6	2.32857	cubit	2ft	of 1.16666ft	Rt Russian	½

Book V

THE MEASURES OF GAUL

THE MEASURES OF GAUL

5.1 OVERVIEW

In a work entitled *Arithmetical Books*, written by Augustus De Morgan in 1847[1], he neatly identified the primary causes of modern confusion regarding the modules of ancient metrology. Although this particular work by De Morgan was a general treatise on the extent of mathematical knowledge at that time, he devoted a few pages to the vagaries of metrology. Due to the recent introduction of the metre, there were few intellectuals of his period who did not, to a greater or lesser extent, make a study of the subject. All of them included comparative examples of the various foot measures that were in use throughout Europe, and there was very little concrete agreement for these basic lengths that should have been fundamental. He was very hands-on in his attempts to rationalise what was commonly accepted as the general rules of English metrology.

Firstly, he looked into the most basic module of the English system, the barleycorn. He acquired many examples of the grain and tested their lengths; he then enquired of his boot maker as to the average length of the anatomical foot that the craftsman kept in his records. The variable barleycorn was very inconclusive regarding any kind of a standard; this echoes the assumption of the present author, that in all nations there was a comparison with subdivisional lengths in terms of local cereal and other grains with which the layman would be familiar. They were merely utilised to give a visual idea of the referred length, in no way were standards ever scientifically based upon them.

However, the length of the foot proved more noteworthy, this is because the boot maker gave an average length for the male foot of 10.26

---

1 De Morgan, Augustus. *Arithmetical Books, sub. From The Invention Of Printing Till The Present Time*, Taylor and Walton, London 1847 pp 6-10

inches; this is the half Root royal Egyptian cubit to within .2 of one per cent. It is the one seventh of the overall height of a man near six feet, as memorialised by Da Vinci. Those who could afford to employ a boot maker would be from the middle and upper echelons of society; the better nourished, more selectively bred, they were therefore above the strict average of all adult males – and more nearly approaching "canonical man." De Morgan listed more of these anatomical feet, which he referred to as *geometrical*, but what has percolated down to the present as the "geometrical" foot is the more rational proposition of Alexis Paucton: that it is the four hundred thousandth part of the geographic degree. In this case it would be the Assyrian foot which is the least foot to achieve a "mathematical" length as opposed to the "anatomical;" Assyrian feet are nine to ten of the Greek.

De Morgan remarked of the many men who since the seventeenth century: "had made a true restoration of the ancient measures," (by which he meant definition): "*There arose a disposition among those who inquired into the subject to seek a mystical origin of weights and measures, on the supposition of some body of exact science once existing, but now seen only in its vestiges; a disposition that is not yet entirely extinct. Some speculated on the pyramids of Egypt and tried to establish that the intention of building those great masses was that a record of measures founded on the most exact principles might exist forever. But more turned their attention to the measurement of the earth, and, by assuming nothing more difficult than the degree of the meridian a thousand times more accurate than that of Eratosthenes was in existence hundreds, if not thousands, of years before him, it was easy enough to make out that the whole system of Greek, Roman, Asiatic, Egyptian &c. measures was a tradition from, or a corruption of, this venerable piece of lost geodesy. There runs through all these national systems a certain resemblance in the measures of length; and, if a bundle of faggots were made of foot rules, one from every nation, ancient and modern, there would not be a very unreasonable difference in the lengths of the sticks.*"

Many a true word is spoken in scorn. That, which De Morgan appears to distain in the opinions of his more speculative contemporaries, is looking more like statement of fact in the light of accurate modern assessment of these antique measures and monuments. For example, Eratosthenes' measure of the dimensions of Egypt and of the world, we now know is not merely close, it is precise once the correct modules, and the evidence given for those modules, are identified. Likewise, it is beyond doubt that certain variations in the measurements are dictated by the variations in the meridian degrees, and yes, great monuments of antiquity do indeed preserve these standards as one of their design functions. There is also, undeniably, a closeness in the length of the "feet of all nations" that goes beyond correspondence in length; in all nations there is equivalence in the nature of the foot as being the basis of all other modules.

This is even true of China, their *chi* that is loosely translated as "foot" more properly means "span" which is the geometrical foot referred to by De Morgan, it is the half cubit or *pes naturalis* – the anatomical foot. All of the foot and span measures of China have equivalence in length to the Eastern and to the European modules. Additionally, the itinerary measures, li, of either 1,500 or 1,800 chi are identical in length to the many European *stadia* and double stadia (*diaulos*). Further to this, these Chinese li being both twice 500 and 600 mathematical feet in length have a further multiplication by ten as the *pu*, and would therefore be equivalent to European 5,000ft itinerary miles and 6,000ft nautical or surveyor's miles.[2] Although in the interests of trade, there were Roman envoys in China and Chinese envoys in Rome, this cannot be how the measurement system became disseminated; because the same modules in concurrent use during the Han dynasty of China and Augustan period of imperial Rome were of great antiquity at that time. These facts regarding both universal modules and the application of these modules clearly indicate a common origin that must reach back millennia.

Modern forensic methods of organic analysis are revealing that at least as long ago as the Neolithic, men were far more peripatetic than they have previously been given credence. They migrated, made pilgrimage, commuted and traded over thousands of miles the world over. Not as aimless hunter-gatherers wandering in circles, but people with purpose who knew where they were going; they had maps or the equivalent, and they had well-established roads and tracks.

Before we investigate these itineraries, it would be as well to identify the foot measures that have come down to us as "Gallic." De Morgan's

---

2 *The Measurements of China*, from vol 2, this series

mention of the closeness in length of the various foot measures that he investigated has been the primary cause of scholarly confusion regarding metrology. Perhaps the most educated man who made a concentrated study of the subject was Flinders Petrie, and as astute as he was, he too confused many quite separate modules together. This observation is most clearly evidenced in his influential contribution on metrology that was included in the *Encyclopaedia Britannica* in several editions from 1911 onward, entitled *Weights and Measures, (Ancient Historical)*. In this essay, he listed many foot and cubit lengths as sub-headings of the various sections.

One can immediately see the primary reason for these errors that he propagated; he used inches to express the lengths under consideration in conjunction with *averaging* the variations. It is his treatment of the common Greek foot, the second in his list, where this is most obvious. He recorded it under the heading 12.45 inches, and gave the following variants as examples of its use: in the Propylaea of Athens — 12 .44ins; at Aegina — 12.4ins; Miletus — 12.51ins; the Olympic course — 12.62ins. He then states that analysis of thirteen buildings in Greece gives an average of 12.45ins; this is followed by examples from Etruria at 12.45 ins and from Roman and medieval England as 12.47ins. According to the general structure of metrology all of the variants in inches would be precisely as follows:

COMMON GREEK	Root Reciprocal	Root	Root Canonical	Root Geographic
	12.27ins	12.34ins	12.41ins	12.48ins
	(31.172cm)	(31.350cm)	(31.530cm)	(31.710cm)
	Standard Reciprocal	Standard	Standard Can	Standard Geog.
	12.30ins	12.37ins	12.44ins	12.51ins
	(31.243cm)	(31.422cm)	(31.602cm)	(31.782cm)

As can be seen, Petrie's values are very closely in accord with the theoretical lengths, except in one very obvious example, the longest; this is the foot of the Olympic course that he gives as 12.62ins, and expressed in inches it may obviously be confused with the other, accurately given, examples. However, the common Greek relates as 36 to 35 of the more familiar "Greek" that is universally acknowledged to be 25 to 24 of the "Roman." The Root value expressed in decimal feet for the Root common Greek is therefore 1.0285714ft. Take the value of 12.62ins that he gives for the foot derived from the 600th part of Olympic stadium and in feet, this is 1.05166ft. Thus expressed in decimal feet, it is obvious that this module has gone beyond what may be classified as common Greek, it is one of the variants of the "Persian" foot. [3] Petrie's given value for *this* foot is accurate to .8 of one percent.

This is one example that we may be quite sure of, because the starting lines are still in situ on this course and have been accurately measured. This illustrates how decimal feet make analysis much simpler – and clarity is even further occluded when using the metric system, than it is with inches. The common Greek foot was also the Rhineland foot and the third of the Scottish *elwand* that was known in Britain throughout the Middle Ages as the *yard and inch*.

This common Greek identification exercise is merely one glaring example of the confusions that can arise in comparative metrology, but it is the many values that are forwarded for the lengths of "Belgic" feet that are most commonly misidentified. A veritable assembly of quite different lengths are often referred to as "Belgic" feet, and once again, it is Petrie's lists that are the most informative as to how these discrepancies occur. Petrie writes with such an air of confidence on matters of metrology that it borders on arrogance. His sweeping testimonials as to derivation and dissemination of the varied modules, makes them sound like statements of fact, whereas it is mere opinion. Concerning the Belgic foot, statements such as *"This measure does not seem to belong to very early times it may probably have originated in Asia Minor." "Thence it passed to Greece."* As to its presence in England *"It is directly in line with migration of the Belgic tribes into Britain."* Such assertions are purely conjectural and the system as a whole is of such antiquity and universality that no such firm conclusions may be forwarded, either to its origins or dissemination.

Petrie gives the examples for the Belgic foot under the heading 13.3 inches and goes on to give values found in various localities that are close to this value. These include – It is found in Asia Minor as 13.35ins, given by Hultsch as 13.1ins, in Greece as 13.36ins, In Romano Africa as average

---

3 See *Greek Stadia Lengths*, this volume.

13.45ins, the palace at Mashita (Persia) 13.22ins. He goes on to connect it with the Belgic foot stating that "*The Belgic foot of the Tungri was legalised by Drusus as one eighth longer than the Roman foot or 13.07ins*" and "*... the small difference from 13.3ins is not worth notice.*" Astonishing! That's nearly six millimetres and would be very noticeable in quite a short distance. It would be over seven centimetres on a twelve foot or pertica measuring rod. The values given in the order as above from the heading 13.3 inches, in feet are 1.108ft, 1.1125ft, 1.09166ft, 1.1133ft, 1.1208ft and 1.10166ft.

Apart from the value tendered by Hultsch at 1.09166ft, none of the others are a Belgic foot. All of the others are variants of the archaic English foot of the *yard and full hand* that was prohibited in England as a legal measurement in 1439. The reason for this abolishment is that the cloth merchants or buyers were replacing the cloth standard of the time, which was the *yard and inch*, with the longer yard and full hand. At the expense of the seller, the yard and inch amounts to a little over 3% commission and buying by the longer measure the merchants increased their revenue to 9%. It took government intervention to stop this practice by abolishing the yard and full hand as a legal value.

In the course of his article Petrie went on to confuse yet another module as the "Belgic" – "*Here we see the Belgic foot passed over to England, and we can fill the gap to a considerable extent from the itinerary measures. It has been shown that the old English mile, at least as far back as the 13th century was of 10 and not 8 furlongs. It was therefore equal to 79,200 inches, and divided decimally into 10 furlongs 100 chains or 1000 fathoms. For the existence of this fathom (half the Belgic pertica) we have the proof of its half, or yard, needing to be repressed by statute in 1439 as "yard and full hand," or about 40 in., – evidently the yard of the most usual old English foot of 13.22, which would be 39.66.*" Clearly, this is not the Belgic; it is the Saxon or Northern foot that was adopted in a statute of Elizabeth. It replaced the natural mile of 5,000 feet as the mile of 5,280 feet, this being 8 furlongs of 660 feet. This was done to bring the itinerary mile into line with the agrarian measure of England, which was the acre. The side of a ten-acre square being one furlong, this furlong is the Saxon 600 feet stade. The acre of 43,560 square feet then becomes the rounded 36,000 square Saxon feet.

Nowhere in all of this is the Belgic foot, yet we know on the best of evidence a very precise definition of this module. This is because it was identified by the governor of Gaul, Nero Claudius Drusus, brother of the emperor Tiberius, who fixed it at nine to eight of the customary Roman foot as the exchange between Rome and Gaul. This would be a constant, because all of the variations of the Roman foot are mirrored in the variations of the Belgic. According to P J Huggins,[4] this comparison is confirmed by the writings of Hyginus Gromaticus. Who stated that he encountered a measure used near Liege in Germania, which was within the territory of the Belgae, that was longer than the Roman foot by 1½ inches, and this is the same 9 parts to 8 of the Roman, as legalised by Drusus.

Huggins too, failed to distinguish between the Belgic and other close measurements stating *"we arrive at the range of lengths for the Drusian foot of 32.96cm to 33.57cm, or between 13 and 13.2 modern inches."* Huggin's 13 inch measure would be exactly 12.989 inches as the Standard Belgic foot, but his 13.2 inch is exactly the Saxon foot which is adapted Sumerian. This observation displays the similarities of the branches, 13.2 Roman inches being the Persian foot, 13.2 English inches being the Saxon foot, and 13.2 common Egyptian inches is the Belgic foot.

A very good example follows showing how these measurements expressed in decimal feet are distinguishable, one from another; with inches these distinctions are far less clear, and when conveyed in centimetres they smear undetectably into each other. Although the misunderstandings that occur when considering these modules at the foot length are perfectly understandable – the dissimilarities become more distinct at the level of their multiples. This largely begins with the five and six multiples, the pace and fathom; at the level of the stadia it is much clearer, whilst at the distances of the mile and league they become quite obvious. This is what is intended to explore, because on the broader canvas of itinerary distances things are far more informative as to both the basic modules and to the ancient skills of accurate surveying.

To summarise these foot lengths that are collectively confused as Saxon, Northern and Belgic, we arrive at the following unit fraction relationships that demonstrate their separate identity; we use the Root

---

4 Huggins, P J. *Anglo Saxon Timber Buildings*, London Archaeologist, Autumn 2005

values of each for these comparisons:

common Greek	1.0285714ft	36/35 of the English (x 3 = *yard-and-inch*)
greater Belgic	1.08ft	9 to 8 of the Roman .96ft
Sumerian	1.0909ft	12 to 11 of the English
Saxon or Northern	1.1ft	11 to 10 English (modified Sumerian)
archaic English	1.111ft	10 to 9 English (*yard and full hand.*)
Nippur	1.125ft	9 to 8 of the English

5.2 The Gallic – Celtic – Roman itineraries

A simplistic rendition of the agrimensores and gromatici at work, surveying a road. The reality would be a scene of frantic activity involving hundreds of men. Note the milestone in the foreground.

Wherever in Europe one drives along a particularly long straight section of road that is not modern, the assumption comes to mind that "This must be a Roman road." Nevertheless, this widely conjectured supposition may be only partially correct; inasmuch that at some point in the road's history, it had probably been resurfaced by them. Excavation of the Roman roads, more often than not, reveals that they have been laid over a pre-existing track. Many of the roads, in terms of straight alignment over very long distances, exceed the accuracy that we know the Romans were capable of. The instruments of the Roman surveyors were basically the groma, chorobates and the graduated decempeda staff. These give the plumb, the level and the elevation measure.

Although near total accuracy can be accomplished over relatively short distances of a few miles by using these instruments, much of the upper limit would be dictated by the nature of the terrain, many of the straight stretches of Roman roads far exceed what was methodically possible for them. The majority of specialists who have written on such matters have been unable to offer a satisfactory explanation. All of the methods, which have been tentatively forwarded as to how it could have been done, are unsatisfactory. From the trial and error of long lines of men constantly to-ing and fro-ing till they come to a semblance of a straight line and back sighting, may be dismissed. So too can beacons of fire in baskets to aid visual sight lines, curvature and many other factors preclude this over very long distances. It has even been suggested that homing pigeons were released and their flight paths noted. These difficulties are most ably documented by Hugh Davies in his *Roads In Roman Britain*, and along with all other authors on the subject, he has had to admit defeat as to a rational explanation.

The only known reference to a method which is feasible, was recorded by Heron of Alexandria, who proposed that an accurate base line be measured and the area be mapped from this line by triangulation. This, of course, is the only known way to do it, but the Romans never used the method. Because by Heron's time of the first century – during the time lag between his discovery and its dissemination, the roads were already built. There is an example from Greece of a sighting device called an alidade, placed upon a horizontal plate that is marked out in angular degrees, and above this table, a movable sighting device that was calibrated in vertical degrees. Add lenses to the sighting device and you have a basic theodolite.

Evidence, as gathered by Robert Temple in his book *The Crystal Sun,* proves that lenses capable of any degree of magnification were abundant in the ancient world. The point being that relatively sophisticated surveying devices were *possible* to construct in antiquity – but the Romans never did it.

The implication of this fact is that the Romans found that the territories in which they operated had already been accurately surveyed at some previous and indeterminate time before their arrival. This came as no surprise to them because the same was true throughout Italy; the fact that ages-old pre existing long distance roads were ubiquitous would therefore, be deemed worthy of little comment. This is not say that the Romans did not build roads from scratch, they built link roads by the thousands; this is a matter of record. These were originally for military use and the army would have among their ranks, men with the necessary skills for any purpose; the highly mobile legions were self-sufficient.

The Royal Engineers, estimating the necessary manpower for the construction of basic tactical roads, conclude that 1,000 men could progress at a rate of over one mile per day. This would entail clear felling of timber and scrub, removal of the overlay, levelling a central carriageway with a timber kerb, providing rudimentary drainage and constructing walkways across marshy areas. The more permanent roads would follow at leisure.[5]

Regarding the Roman roads in Britain, there are a number of inexplicable features that indicate a pre Roman foundation. The diagram opposite is the route of Stane Street; it is the Roman road connecting Londinium to Noviomagus Reginorum – the modern London Bridge to Chichester; some eighty miles (Hucker). The first thirteen miles, London to Ewell, is perfectly straight, it then deviates into a series of straight stretches before homing in to its intended destination. What is peculiar about this arrangement is that the original thirteen miles configuration of the road is exactly aligned upon this destination. The implication being that the engineers knew exactly which direction to take; how?

---

5 Hucker, Richard Adrian. *How Did the Romans Achieve Straight Roads*

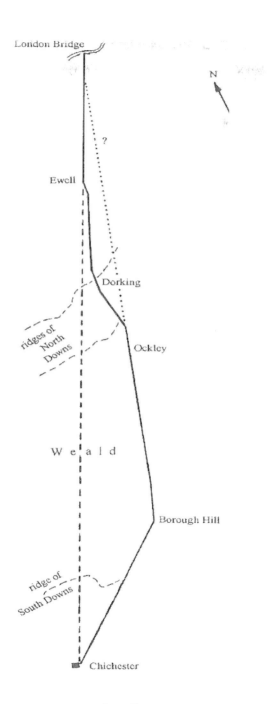

Stane Street

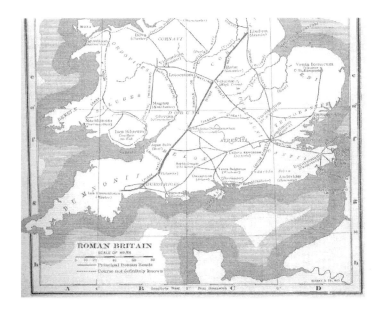

The Fosse Way

Another example is the Fosse Way, a major road of some 250 miles connecting Exeter to Lincoln. Along this route, in the 182 miles between Ilchester and Lincoln it deviates by no more than six miles from a direct line, on its course, there are several straight stretches of around eighteen miles; these would be double "stages" each of 20 Roman miles. Once again, the question is posed– how did the engineers know the direction of the destination so accurately?

There is further significance to this street regarding the inexplicable accuracy of the Roman agrimensores; the point where the Fosse Way crosses Watling Street on the Warwickshire-Leicestershire border, was reputed to be the geographic centre of England. In 1712 a high cross was erected at this site bearing the inscription "*If, traveller, you search for the footsteps of the ancient Romans, here you may behold them. For here their most celebrated ways, crossing one another, extend to the utmost boundaries of Britain; here the Vennones kept their quarters; and at the distance of one mile from hence, Claudius, a certain commander of a cohort, seems to have had a camp, towards the street, and towards the foss a tomb.*" This point is where three counties meet and is most accurately the geographic centre of

England; nearby was the Celtic omphalos of the territory that comprises England, in the shape of Croft Hill, some 420 feet in height and long used as a beacon site.

Stukeley's drawing of High Cross. It was erected to replace an ancient oak that had marked middle England from the time of the Romans, struck by lightning soon after its erection its reconstituted remains now stand beside the road.

Much more evidence of a similar nature could be presented to substantiate that carefully surveyed features of Celtic Britain long pre dated the Roman presence. The principle solid evidence that corroborates this statement is that the identical straight roads, connecting towns and sacred centres are evident throughout the Irish landscape, yet the Romans never visited Ireland. Furthermore, the ancient and traditional "Centre of Ireland" is in Westmeath on the Hill of Uisneach that is marked by a great

Ail na Mirean, The Stone of Divisions

stone. This stone is where the borders of Ireland's five provinces, Leinster, Munster, Connacht, Ulster and Mide meet.

Once again, the levels of accuracy as to the precise centrality of the geographic landmass of the whole of Ireland are phenomenal. There is hole that is bored into the top of the Ail na Mirean stone (*shown above*), whose most probable purpose was as the socket of a pole that could be used for sightlines. John Michell most ably deals with the evidence for such national practises as locating the geographic centres of their administrative territories in his book "*At the Centre of the World.*"

These divergences into Britain from Gaul are no digression, the cultural and familial ties that bound virtually the whole of Europe together as "Celtic" is self-evident. The Celtic world stretched from the Atlantic to the Black Sea and tribes from many indirectly related sources had peopled this territory; there is no Celtic genetic strain. Descriptively, one man's Celt may be described as dark, stocky and brachycephalic – Irish-Breton; another as tall, muscular, blonde and dolichocephalic – Germanic. The continent may be fairly described as a federation of tribal nations, much as North America before western invasion. Indeed, archaeologists have often advanced this analogy; Aubrey Burl in particular, has repeatedly

Celtic, left, and Native American, right, illustrative of cultural similarity

forwarded Neolithic-Bronze Age tribal and cultural parallels with Native Americans and much could be understood of the Celtic tribal customs through the recorded history of the American Indian.

Cultural similarities between North America and ancient Europe go far beyond these superficial resemblances. Earthen mounds and henges abound in North America, great timber circle lodges for councils, tribal ritual and meetings at permanently established centres are common to both. Although fragmentary, evidence of carefully planned and accurately aligned roads throughout America, both north and south is accumulating. One example is the Great Hopewell Road that stretches unerringly for 60 miles between Chillicothe and Newark; there is a carefully planned network of roads surrounding Chaco Canyon and "Indian trails" connect all of North America. Most importantly, it is becoming obvious that the natives of the advanced cultures of Mexico (who were directly related to the Hopewell and Adena of Ohio) had used a measurement system that was identical to the Spanish, who supplanted them.[6]

Both cultures were aware of the extent of their vast territories and the inhabitants therein. Many languages were spoken throughout these areas and in America there existed a universal sign language for communication. If Julius Caesar is to be believed, the Celts communicated through the

---

6   John E Clark Aztec *Dimensions of Holiness in The Archaeology of Measurement, Comprehending Heaven, Earth and Time in Ancient Societies.* Edited By Iain Morley and Colin Renfrew. Cambridge University Press 2010

medium of Greek notation, although this was probably used merely for bureaucratic purposes because it is well known that their laws, customs, rituals, cosmologies, esoteric teachings and history were committed to memory and never written down. However, Greek was in all probability the lingua franca of the Celts.

The map below illustrates Northern Gaul at the time of the Roman invasion; the proximity of Britain and Gaul is a great deal more than geographic. The Belgae and the Atrebates of Gallia Belgica had extensive territories in Britain. Silchester, in Hampshire south of Reading, is the Roman Calleva Atrebatum, and was founded upon the Iron Age capital of the Atrebates. The Belgae are located by some historians in eastern Dorset, and by others in western Wiltshire.

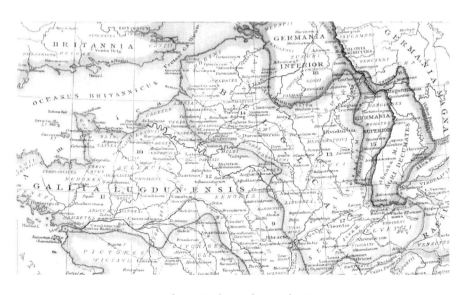

Northern Gaul according to the Romans

There are two principal itinerary accounts, through many copyists, of the extent of the Roman Empire. The Antonine Itinerary was in the form of a bulletin or register of the positions of towns and military forts along the extensive road system and the distances between them; the oldest surviving copy from Rome is from the time of Diocletian, early fourth century and the original imperial patron was probably Caracalla, during the early third century.

The other, more extensive register has come down to us in the form of a map, as the Tabula Peutingeriana, anglicised as the Peutinger Table. This is believed to have its origins as being commissioned by Julius Caesar, but interrupted by his assassination, it was completed by Marcus Vipsanius Agrippa, on behalf of Augustus. The first localised maps from the field were etched onto bronze, probably for reproduction through rubbings, and the full version was engraved on marble and placed in Rome in the Porticus Vipsania on the via Lata. It was subject to many revisions until the fifth century and the faithful copy that has come down the ages was by the sixteenth century scholar-antiquarian Konrad Peutinger, that he reproduced from a thirteenth century copy.

In the Antonine Itinerary, all of the distances are recorded in miles; and in the Tabula as leagues of 1½ miles. The Antonine Itinerary has been repeatedly criticised by historians on account of its distance discrepancies. Unversed in metrology, the majority of these critics did not realise that all the distances were not necessarily given in the same Roman "mile" unit. They remark admiringly, on the one hand, how *accurate* they were, yet on the other hand, when it comes to the recorded distances between points connected by the roads, remark how *imprecise* they were. A careful analysis of these recorded itinerary distances provides as much confirmation for a pre-Roman origin of the system as does the physical evidence. Just a smattering of these distances that are applied to recognisable place names in Britain, will be given as possible alternatives of the Roman mile.

If the distances are carefully plotted on the map and the number of "miles" given in the Antonine Itinerary is divided into these physical lengths, other known measurements for the mile fit these distances with remarkable regularity and accuracy. The Itinerary comprises fifteen sections, each listing a number of connected towns or settlements.

The first example of this is in Itinerary I, which begins at Hadrian's Wall. From High Rochester, Northumberland, the distance given to Corbridge is 20 miles, but consultation of the map gives 23.04 statute miles and this is exactly 20 nautical miles of 6082.56 feet (the Admiralty value for Britain). This nautical mile is 6,000 Greek feet of 1.01376ft or 10 stadia, thus the distance is 200 stadia overall. (The geographic mile is also the measure used in the itinerary 13, from Canterbury to Dover).

Then from Corbridge to Ebchester, Durham, the Itinerary gives

9 miles, whereas the actual distance is 9.6 statute miles. Not only is this exactly 9 x 5,000ft miles of 1.1264ft, which is a variant of the ancient British foot of the "yard and full hand" (archaic English), it is also 10 Greek miles of 5,000 feet each of 1.01376ft. It is not until the distance given from Catterick to Aldborough (20¼ statute miles), stated to be 22 miles, that we have a recognisable Roman mile of 5,000ft of 0.9732096ft.

Examples of the 5,000 English feet mile are exactly the 16 miles given between Wall, Staffordshire to Mancetter, Warwickshire; this is also the solution to the 17 miles from Aldborough to York. An example of the 5,280ft mile is the distance given from Chester to Tilston, listed in the itinerary as 10 miles, which it exactly is, 10 statute miles. All of these distances are the straight line, or survey distances. Much more of the same nature could confirm theses observations, but only a selection of the places that are immediately identifiable from the Roman place names are listed here.

The distances in Gaul from Peutinger's table, given in leagues, have aroused similar questions as to the accuracy of the Roman reports; and similar comparisons of the measured distances reveal that they are quite correct in terms of related modules. The "Great Gallic League" as the itinerary distance measure of Gaul has a great number of interpretations as to its identity. Petrie states of it "*The Gallic leuga, or league, is a different unit, being 1.59 British miles by the very concordant itinerary of the Bordeaux pilgrim. This appears to be the great Celtic measure, as opposed to the old English, or Germanic mile.*"

The now anonymous native of Bordeaux who made the pilgrimage from northern Italy to the Holy Land in the years 333 and 334, kept a dry-as-dust account of all of the stages of his journey (he was no Chaucer). He recorded the names of the stopovers and the distances separating them along the Danube valley from northern Italy to Constantinople, and from there through Asia Minor and Syria to Jerusalem. It is believed that the length referred to by Petrie was calculated from the actual distance travelled and collated with the account of the Bordeaux pilgrim. 1.59 miles is 7,500 feet of 1.11936ft, the closest value to this is the Standard Canonical Archaic English foot of 1.12ft, this is directly related to the measure of the Antonine itinerary, Corbridge to Ebchester, and is one of the foot values listed above as being confused with the Belgic foot.

By far the most informative person in recent years regarding the itinerary distances of Gaul is Jacques Dassié, an archaeologist who is something of a pioneer regarding the methods of accurate identification of these itinerary distances. He makes extensive use of aerial photography in conjunction with global positioning satellites, then verifies these distances on large-scale maps, making all due allowances to achieve accuracy. He has proven that the Roman milliary stone inscriptions relating to these itinerary distances are not all recorded in terms of the Roman league. He has accurately identified a variety of distances that were termed "leagues" because they have been repetitively found in the course of many hundreds of comparisons.

In an article entitled *La Grande Lieue Gauloise*[7] he uses many widely researched references from older specialists in the field to substantiate the lengths that have been historically proposed as Gallic leagues. Among them were Jean Baptiste Bourguignon d'Anville, a cartographer who was influential in improving the techniques of map-making; who, in 1760, calculated from the distances between the cities of Gaul a Roman league that equates to 2,211 metres. The Standard Canonical value of the Roman foot is .96768ft and 7,500 of them equal 2,212 metres.

Another was Félix-François Le Royer de La Sauvagère, a nobleman, antiquarian, and engineering military officer, he was officially chief engineer for a number of city fortifications. He was an authority on Roman masonry and architecture; in 1770, he calculated a value of 2,225 metres for the Roman league and 7,500 feet of the Standard Geographic classification, of .9732096ft, is 2,224.75 metres. There is much to be said in favour of using this Standard Geographic classification at this latitude as it is often found in the itinerary modules in Britain; for example it is the same Roman foot basis of the 22 Roman miles between Catterick and Aldborough, as listed in the Antonine itinerary.

Perhaps the most interesting length for the Gallic League quoted by Dassié is that proposed by Theodore Pistollet De Saint-Ferjeux; he was a noted specialist on such matters being the author of many antiquarian books, including *Memoire Sur L'ancienne Lieue Gauloise*, published in

---

7   Dassié, Jacques. *La grande lieue gauloise [Approche méthodologique de la métrique des voies]* Gallia, 1999 vol. 56 no.1 pp 285-311

1852. He stated that the Gallic league was 2,415 metres, and one and a half English miles is 2,414 metres. This means that the *pied de roi* of the Franks long predated its generally accepted introduction into France. This was supposed to have been the year 785 when Charlemagne adopted the *hashimi cubit* of the Arabian empire of Harun al Raschid as the pre-eminent unit of the Franks; this was done as a token of the peaceful alliance between the erstwhile Christian and Islamic enemies. The foot of Charlemagne was a variant of this foot of the English mile of 1.056ft.

Two other authors are quoted by Dassié, Auguste-François Lièvre, librarian and archivist of Poitiers, author of "*Les chemins gaulois et romains entre la Loire et la Gironde. Les limites des cités. La lieue gauloise.*" 1891. The other was Auguste Aurés, who wrote "*Note sur le système métrique des gaulois*" 1866, both are mentioned together because they proposed the identical length for the Gallic league.

Aurés, the author of 14 memoirs concerning the Gallic league, in 1865 proposed a value of 2,436 metres. Liévre later confirmed this distance in 1893; he described the methods he used in this thorough determination in following the route from Tours to Poitiers, the distance in a number of stages was 102.3 kilometres, which was 42 leagues according to the Tabula Peutingeriana, giving the same length for the league as Aurés' 2,436 metres. This is within 2½ metres in the league of the original reckoning of the pied de roi adopted by Charlemagne. It is directly related as 1.008 to 1 of the Persian foot reckoning of De Saint-Ferjeux as previously given, underscoring that the pied de roi – Hashimi cubit were anciently used in France long before its adoption by Charlemagne.

The term *hashimi* in Arabic means royal, and throughout the centuries, many cubits of quite different derivations have been termed hashimi. This particular cubit was provenanced by Skinner[8] who was particularly astute in module identification; he noted the hashimi cubit (of 640mm), was a variant of the Persian cubit of Darius the Great, which it is. The foot of the Persian cubit of Darius is therefore 1.05ft, the foot measure of the league proposed by Pistollet De Saint-Ferjeux is 1.056ft and the hashimi foot - original pied de roi - is 1.064448ft; this a related series of the same

---

8 F.G.Skinner, *Weights and Measures. Their ancient origins and development in Great Britain up to 1855*, HMSO 1967

foot. One of the ancient Gallic leagues is therefore half of the English three-mile league, which from a number of sources, is also the length of the Persian *farsakh* or *parasang*.

The Parasang, that is also the three-mile English league, was calculated to be the distance to be walked purposefully by a man in one hour, while the reckoning of a parasang in Arabia was taken to be the walking speed of one hour for a horse. A parasang was divided into 30 stadia, these would be 500 feet stadia of exactly one tenth of an English mile, the constituent foot – 1.056ft. This foot was widely used by the Greeks and the 600 feet stadium of Olympia (192.5m) is the foot of the Standard classification, 1.05238ft. Thus, it is clearly demonstrated that the identical foot-measure was widely used and the variations that are detectable, are regularly found and therefore deliberate.

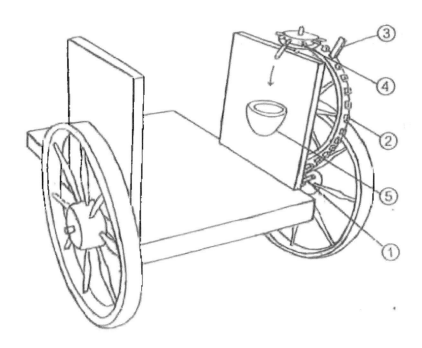

A Roman Odometer similar to that described by Vitruvius

The odometer in the diagram on the previous page consists of: **1**, a drum wheel affixed to the axle with one tooth projecting; **2**, a gear wheel with 40 teeth; **3**, one tooth projecting above the others; **4**, a revolving dish with ten teeth; **5**, a metal dish into which the balls audibly fall.

From the above description, with one revolution of the carriage wheel it moves the gear wheel one tooth. After 40 revolutions, the gear wheel turns the horizontal dish one tenth of a revolution. After 400 revolutions of the carriage wheel a ball drops into the metal dish signifying that one mile has been traveled. (In Vitruvius' description the gear wheel has 400 teeth, and the horizontal dish but one.)

It was Vitruvius' description of this odometer that first alerted the author to the practical use of the 175 to 176 fraction that is common to all branches of metrology. This is because Vitruvius stated that the carriage wheel was four feet in diameter and after four hundred revolutions, it would have travelled one mile. This clearly indicates that his pi ratio was 3.125; this calculation was common in the ancient world, it is also recorded from Mesopotamia, but in all instances the diameter is four or a multiple. This means that it was only used to simplify calculations in whole numbers, the actual circumference, or distance traveled, would be in a module the 175th part greater because the difference between 3.125 and the accurate 3.142857 is 175 to 176.

By using the four roman feet diameter carriage wheel one can measure the distance travelled in any of the mile measures by adjusting the number of teeth on the gear wheel. For example, if there were 44 teeth then the distance covered would be the Persian- English mile that are common to the Gallic leagues. This is a very practical device particularly when allied to horse speed. The following calculations are taken from an old English journal.[9] 2 ½ miles an hour for strong draft horses up to 7 ½ miles an hour for horse teams with heavy carts. Mail or messenger horses go at 10 to 12 miles an hour for two such stages a day. Thus, Romans and Celts could accurately know distance or potential distance travelled. The odometer would be handy for rough estimates on surveys or accurate records of distance on completed roads.

The foot measures of Belgic tribes that we have demonstrated to be

9 *The Quarterly Journal of Agriculture* volume iv 1884 William Blackwood, Edinburgh

classifiably quite distinct values other than "Belgic," have a clear relationship with another Gallic league as recorded by Dassié. This additional value of the Gallic league is one of 2,490m, and 7,500 Belgic feet of 1.08864ft is 2488.6m; this is almost the sole value of the Gallic league in "Belgic" feet. One other was calculated in 1970 by Daniel Jalmain in a presentation at an international seminar on *Cartographie Archéologique.* He found the Gallic league to be 2,450m.

This is a doubly interesting conclusion, on the grounds that its basic foot is a well provenanced measurement that is constantly misidentified and its explanation pertaining to the structure of metrology is most informative. In exact terms this Gallic league would be 2,449.3m that is within .7 of a metre of Jalmain's finding of 2,450m. The basic foot is 15/14 of an English foot, 1.0714285ft; this exact length was classified by William Bell Dinsmoor as a "Doric" foot, but it would be correctly termed as a "lesser Belgic foot." Increased by 1.008 it is the Belgic foot of 1.08ft; this in turn is 9 to 8 of the Roman foot of .96ft. This lesser Belgic of 1.0714285ft is then 9 to 8 of the lesser Roman, which would be .95238ft. Not only is this 20 to 21 of the English foot it is the ancient reckoning at 5 to 6 of the royal Egyptian foot of 1.142857ft (Heron), which in turn is 8 to 7 of the English foot. Complex, but not in the slightest complicated – and exact.

Jacques Dassié is what may be termed a progressive thinker in his approach to these distance analyses; the majority of the notoriously conservative archaeological community would as soon ignore these variations in the leagues and ascribe them to accidental inaccuracies. Of recent years another academic antiquarian has emerged who has approached the Celtic culture with uniquely perceptive insights. This is Graham Robb who presented his findings in "*The Ancient Paths*" subheaded - "*Discovering the Lost Map of Celtic Europe.*" The book aroused much popular critical acclaim and considering the innovative ideas that it presented, its persuasive presentation preserved it from the outright dismissal usually reserved for revolutionary "discoveries." The academic press were not so kind.

The author was delighted with the work because the findings support the conclusions that had been reached, as set out here, from the quite different evidential perspective of metrology. This includes the network of durable and carefully surveyed system of roads that were so obviously pre-Roman. Much of Robb's interpretation of the evidence relies on place names, and these are often the best substantiation for theories; place names are often descriptive of function and adhere to sites for millennia, they even survive changes of language with slight pronunciation alterations. He uses Caesar's *Gallic Wars* for much of his evidence, but merely in order to point out that he lied by omission. He calls the book "self serving" in as much that Caesar did not mention the sophisticated infrastructure of the Celtic landscape that he so conveniently took advantage of.

This took, not just the form of existing high quality roads that he utilised, but also the many strategic bridges that were long in place wherever he went – and took the credit for their construction. In Robb's words; *"Otherwise one would have to believe that the Gauls gave the name 'briva' to about thirty different towns (Brienne, Brioude, Brive, etc) and then waited for the Romans to come and build the eponymous bridges."* As to the existence of high quality roads he remarked that the marching speed of the legions was far above that of the average for the Roman empire; the figures, of course, were adequately chronicled in the Gallic Wars, that recorded what Caesar did, but not how he was able to do it.

He was able to march thousands of men from Sens to Orleans in four days, twenty Roman miles a day. Caesar himself was able to reach Belgic Gaul from somewhere south of the Alps '*in about fifteen days*', which is at least twenty-six miles a day. Robb claims that the Gauls were even faster, a battle weary army of one hundred and thirty thousand men marched from Bibracte to the territory of Lingones in just four days, some thirty miles a day – '*which gives some idea of the resilience of the road surface.*'

As well as the material infrastructure of the roads, there were complete itineraries of the network: *"In seven years of campaigning Caesar always knew when he was entering or leaving a tribal territory, and – most telling of all – he always knew the exact distances to be covered. Distances in Gaul were so accurately measured and so comprehensively plotted that even after the conquest, the Romans continued to use a standardised Gaulish league instead of the Roman mile."*

At this point, we shall briefly leave Robb's Gaul in order to give corroboratory evidence of the preceding, concerning the road system in Britain. As previously stated, many researchers in the past have remarked on the existence of a vestigial pre-Roman network of good quality roads;

but heretofore, such commentary has been almost apologetically guarded, mentioned as an aside rather than a bold statement of fact. Such reticence seems now, thankfully, to be diminishing with a fresh generation of archaeologists and academics, such as Robb, who have the confidence to present what the evidence suggests and to shrug off the reactionary caution that has hampered innovative progress.

The findings of M C Bishop,[10] a specialist in Roman military affairs, without going too deeply into his research, because his credibility is outstanding; the points he makes in his – *The Secret History of the Roman Roads of Britain* is fully supportive of the theory of advanced technology regarding the accuracy of Celtic surveying. In his words: "*Our secret history of the Roman roads of Britain can thus be summarised as a series of 'secrets' that have been overlooked or underplayed, rather than suppressed; the wise reader should always be suspicious of any book that claims the archaeologists have indulged in a conspiracy.*"

"*We have seen how wheeled traffic, and the roads which it required in order to be of use existed long before the Romans came*" – additionally – "*The invading army certainly revolutionised the road network, there is no doubt about that, but they did not invent it. What we perceive as the Roman network also seems not to be the whole story, not just because research has failed to reveal all of the formalised roads in use, but also because much of the prehistoric network appears to have continued in use at the same time. So our first 'secret' is that the Romans did not give us a new road network, they merely adapted an existing one.*"

What one gathers from the theories of Robb regarding the Celtic roads, is that there was a very different motivation for building them other than the purely utilitarian approach of the Roman. In a sense, the roads ritualised the landscape. They were aligned to astronomical phenomena, principally the direction of sunrise and sunset at calendrically significant dates; the solstices, the equinoxes and Beltane in particular. Robb mentions long distance and well-established roads such as the "Heraklean Way" that he claims were aligned on the Mayday sunrise.

The *oppida*, Celtic strongholds, were erected within this astronomical

---

10 Bishop M. C. *The Secret History of the Roman Roads of Britain*, Pen and Sword Books Ltd. 2014

landscape grid at significant nodal points. Once again, the principal evidence for these carefully sited centres of population is preserved in the place names, the first and foremost of which is *Mediolanum* with dozens of related words evolved from this source name, meaning "centre." So similar is Robb's view of the nature of this vast scheme of landscape geometry to that of John Michell, that he may be suspected of plagiarism; but there is no hint of that.

Robb's fresh insightful recognition of the same vision merely serves to legitimise Michell's appreciation of the same scheme that he had recognised some thirty years—and in some significant instances sixty years —prior to Robb. It is surprising that Michell receives no mention his works. Perhaps he has enough on his plate with his acknowledgement of "Druids."

The best founded of the examples of the parallel disclosures of the two men is the existence of the St. Michael's line in England. This is a continuous straight line that connects Lands End in Cornwall to the North Sea in Suffolk – via an alignment of hilltop Churches dedicated to St Michael and St George; along this route are many highly significant Neolithic sites, such as Avebury. Proposed and mapped by John Michell as early as 1967, this is no speculative line of the weekend ley hunter; its firm existence is verified by the fact that it is an extension of the Icknield way, an ancient track that was resurfaced by the Romans and extended from East Anglia far down into Wiltshire.

A path connecting Suffolk to Cornwall has long been enshrined in folklore, and the ancient Celtic king Belinus is accredited with drawing such a line on the map (Michell). But the principal connectivity of the Michael line to the findings of Robb is its orientation; it aligns on the Mayday sunrise, as does the Via Heraklea of Gaul. This is doubly significant because the holy day of St Michael is May 8th and this would be virtually indistinguishable as a solar alignment, from the Mayday orientation.

Not only were the oppida carefully sited along these, essentially calendrical, alignments; but all the sites of their Roman battles are along the same configurations. The inference that one may possibly take from this is that warfare for the Celts was a highly ritualised affair, requiring an auspicious place and date for their battles. These conditions may have been more able to be dictated by Celts in the earlier stages of their seven-years war, but such diplomatic war had ceased to be with the increasing Roman triumphs. The Romans changed the rules of warfare and settled for nothing less than total subjugation of an entire race and annexation of their land. This was not quite how the Gauls regarded the art of warfare – as motivated by rapacity and pitiless greed. They were overcome by a force and determination that was beyond their experience or understanding.

In Robb's work, there is little mention of measurements, either of the modules being used, or of the league distances between any significant point spacing. The exception is the Map, below, it is of a common centre, Molliens Dreuil, to six other towns around this collective centre that are very close to being a common radius distant. He claims the place names are all derived from Mediolanum, and gives the relevant lengths of the various radii in kilometres that are accurate within 100 metres.

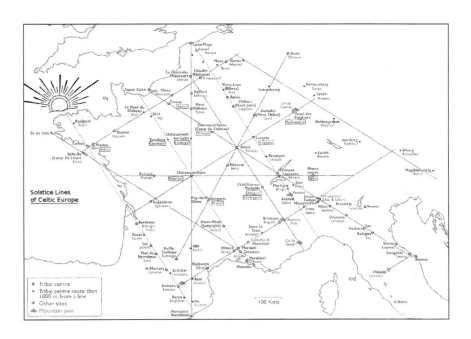

Graham Robb's proposed solstice lines of Gaul

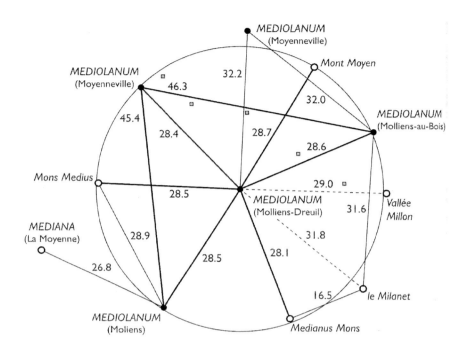

An arrangement of sites in Picardy around a common centre

The six towns that loosely conform to this common radius do so between the parameters of 29km and 28.4km. Two towns visibly exceed this upper limit, at 32.2km and 38.1km, and one radius is less than the bulk distance at 28.1km. For the majority of people who would consider such data, their first impulse would be to average the distances; but this invariably eradicates the detail and any semblance of accuracy.

There are two ways to approach such data sets to identify quanta. The first, because we are dealing with a circle, is to divide the radius by seven. The other is to look for a "customary" unit in the context of the problem. In both cases, the distances must first be converted to feet, then if either method turns up a recognisable unit it may be immediately classified as to the foot variant. Begin with a repeated number, in this case 28.5km; this is 93,504ft – this divided by seven equals 13,357ft. The closest rational number to this is 12,000, divide by this, it is 1.11314ft. This can only be the archaic English foot of the *yard and full hand*. One possible design module is therefore the double 600ft stadia (*dioulos*) of this foot, there would be exactly 84,000 such feet in 28.5km, the closest proven length of the constituent foot would make a total of 28.512km and this is within 12 metres of Robb's data.

If it were in terms of a Gallic league or 7,500 such feet, then the league would be 2,545m. This (undoubted league) has not been encountered, although the *mile* in terms of this foot *has* been identified in the Antonine itinerary. The second way to approach module identification from Robb's data; is to apply a customary measure, in this case, a 7,500ft league. Take the radius measurement of 93,504ft and divide by 7,500 and this equals 12.4672, again the closest rational number is 12. Divide by this, it equals 1.038933ft and once again, this can *only* be a common Greek foot that would be the basic module of this league; the league of 7,500 of these feet would be 2,351m and although it is undoubtedly a *league*, this variant has not been encountered in the Gallic records. Both derived units would be valid identifications; therefore entirely possibly intentional.

These interpretations, although they yield a recognisable module in rational numbers, could not be dogmatically forwarded as a solution; it is more an exercise in module detection and without corroboratory evidence of historical record, must remain just that. Although it must be pointed out, that this tentative solution would be entirely typical of ancient metrological practice. These Celtic oppida of Picardy, because they are arranged to conform to landscape features, permit the hand of man but little choice in these dimensional relationships. When this is the case the module is selected by the agrimensores to be the most rational, and therefore memorable, record of the distance.

This custom is pointed out in the appendix to the Greek Stadia article concerning the length of the Pharos lighthouse causeway (*see p. 244*), although the engineers had no control as to how long this distance was, it was termed the Heptastadion because the length was seven 500ft stadia of the common Egyptian foot. The same convention governed the length of Hezekiah's or the Siloam tunnel of Jerusalem. Stated in a contemporary inscription to be 12,000 cubits in length, the measured length is 12,000 cubits, also of the common Egyptian foot. In both cases because the distances were arbitrary, a module was chosen retrospectively to give the best numerically rational solution.

One of Robb's distances on his map of the section of Picardy is related to the most significant of the acknowledged Gallic leagues; it is the maximum distance of the related oppida from the centre, this is Moyennville and is given as 32.2km. When converted to feet the solution to this distance is immediately self-evident. The distance is 105,643.13ft and is obviously one hundred thousand Persian feet of 1.056ft, or 32.187km, accurate to well within Robb's error margin of 100 metres. This is 20 English 5,280ft or Persian 5,000ft miles and although it is divisible neither by English nor Gallic leagues, it uses the same basic foot as do the leagues proposed by Aurés, De Saint-Ferjeux and Lièvre, and is present at this exact mile length in the Antonine itinerary of England.

Two observations regarding metrology are brought to the fore from these examples; the first is the dangers inherent in the widespread practice of averaging ancient units because all of the solutions may be spoilt. The second is the indiscernibility as to module that is induced by use of the metric expression of the overall length such as 32.2km, which is *impossible* to analyse without conversion.

5.3 DRUIDS

Knowing that "Druid," due to its rather dreamy modern connotations, is a thorny issue to be raised by any historian, Robb, aware of the danger, grasps the nettle of the subject with considerable aplomb. As with the Celtic psyche, which he has an instinctive understanding of, he sees through all the negative associations of Druidry to the essence of their function in Celtic society.

So great was their influence within their tribes, they had the power and authority even to stop battles that had commenced (Diadorus). In all of their artwork and fashion, Robb sees the inspiration drawn from the geometry inherent in nature, from the oak and the seemingly tangled patterns of growth, knowledge of animism and sacred space.

There was a master plan to their endeavours – that was continent wide in vision, extending far beyond the Greek oikoumene, of which they were aware, and so included Gaul into a wider world through an extension of the same Greek geometry. The closest parallel to the Druids as an organisation were the mystery schools of Pythagoras, both were based upon number, number symbolism and geometry; so close were they in philosophy and practise that they may not have had separate origins. Greek writers, Pomponius Mela, Posidonious, Sotion of Alexandria among their number, as terse as they are on such matters, have speculated on the Druid connection to far wider cultural links. Diogenes Laertius likened them to the Persian Magi, the Assyrian and Babylonian priesthood and, further afield, to the Gymosophistae of India.

A Celtic coin depicting the equality of women both as
Druidesses and the more butch among them as fighters

As early as the fourth century BC, Robb claims that the Druids brought their landscape geometry to perfection by establishing "prime" Mediolanum, the first of which was Biturigum, the modern Chateaumeillant, which he claims was the sacred centre of Gaul of the Bituriges. All of the oppida were placed along the solstice lines of the Via

Heraklea. There were three solstice lines termed Via Heraklea, evenly spaced and parallel that were mirrored in the solstice sunset lines that went in the opposite direction; thus forming a grid that were the bases of their tribal land divisions and regions – and governed the choice of their sacred centres throughout Gaul.

Overall, Robb displays knowledge of the tribal interrelationships, their leaders, the role of the Druids, their deeds and geography that is second to none. Nevertheless, certain of his assumptions must be challenged, without overly affecting his conclusions. The first is the nature of the "solstice lines" of the Heraklean ways. At such an angle over such great distances, the direct line could not maintain the straightness that he claims. The further the line travels in a northward direction, the more degrees of latitude it would cross; the more north you are the more the solstice line would have to deviate from a straight line in order to still point to the sunrise. However, he may have inadvertently answered this criticism in the course of his work. This is because he pointed out that the solstice angle at its more southerly source line from the Atlantic to the Alps, is coincidentally half pi on a triangle that was 11 units along its base side to 7 units on the side.

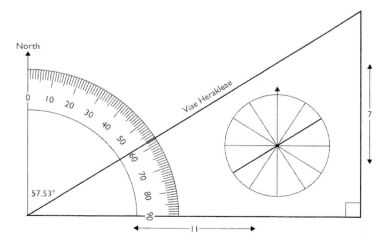

Graham Robb's summer sunrise angle of 57.53° (or half pi) along the Via Heraklea

As Robb continually states in *The Ancient Paths*, the road and Mediolanum sites rely heavily on Pythagorean geometry, then his above pi angle geometry may be the more rational explanation for the repetitive alignment angles. As the pi ratio, is most certainly, intrinsically bound up with metrological values and distances, he may well have answered his critics concerning perceived orientation anomalies.

He also displays a rather touching faith in the efficiency of the groma as a tool of the surveyor, in his belief that the Romans were as competent as their prehistoric predecessors. He claims *"Devices for measuring angles, such as the dioptra and protractor, developed in the third century BC, were used primarily by astronomers. Terrestrial surveyors rarely bothered with such refinements. The protohistoric positioning unit was the groma – a vertical stick with horizontal cross bars and plumb lines which made it possible to draw straight lines and right angles."* After roughly describing the process of the straight alignment, he continues *"the operation could be repeated ad libitum as the survey progressed. Over long distances errors would be averaged out and cancelled."*

If only it were that easy, errors could and did frequently occur and equally likely as cancelling out, they could as often be accumulative. It has been proven to every scientific criterion that the groma is totally unsuited to very long distance accuracy. This deepens the mystery of straight lines being, historically and prehistorically, found the world over, spanning great and geographical distances. We know without doubt that this was accomplished and we also know that the only way of achieving such accuracy is to measure, as long as possible – a base line; from this accurately known datum, one then triangulates to salient landscape features, constantly calculating these great triangles through their angular relationships by means of the Pythagorean theorem. Nevertheless, the Romans never did this.

Robb gives an example of a long distance Roman feat of accurate surveying in the fortified frontier line called Limes Germanicus that ran for eighty kilometres through the hills of the Swabian-Franconian forest; on a twenty-nine kilometre stretch, the directional error was less than two metres. The question must be posed – did the Romans survey this line? Or was it pre-existing as he, and others, have claimed of the roads of Gaul – was this too, merely adapted by the Romans? Other even greater, up to 200km, straight "Roman" roads are known and all men of a scientific background who witness them, express incredulity about Roman abilities to accomplish such accuracy with the groma.

Another instance of surveying on the grand scale in the historical period was mentioned by Robb as evidential. This was the survey of the Chinese Buddhist monk I-Hsing[11] in the early eighth century AD. Now termed Yi Xing, his survey was a notable accomplishment for its sheer scale. "Survey" is perhaps not the right word for the exercise; Yi Xing was responsible for setting up a chain of gnomons, they were largely set up upon a line that stretched from Li An, in Vietnam, along the prime meridian of China to the shores of Lake Baikal at its northern extremity. This distance is in the region of 34 geographic degrees – a distance that would connect the last Druid stronghold of Mona, on Anglesey, to the Great Pyramid. There is absolutely no significance in this; they are merely two notable points.

The fact that this was not a detailed survey is that the examination of Yi Xing took only two years, between 724 and 5. Under the supervision of Nankung-Yueh, the Astronomer Royal, this ambitious program was designed to improve eclipse prediction, reform the calendar and correct the now discredited *cun-qian-li* (*inch-thousand-li*) rule through measuring the shadow lengths cast by an eight feet gnomon at different latitudes. These readings were taken at both solstices and equinoxes; they measured the shadow lengths and the elevation of the Pole Star. The shadows would give the precise north orientation allowing the progression of stations on a northerly meridian. There is no firm agreement on the number of stations that were set up by Yi Xing, and some of them were set up on an east-west line in Henan province; the traditional heartland of China from whence all measures were defined.

The inch-thousand-mile law that had bedeviled Chinese cartography for centuries was the belief that the midsummer noon shadow cast by an eight-feet high gnomon, (*gui biao*) increases by one cun or inch for every thousand li to the north. This obviously incorrect equation came about through the Chinese world-view of a flat earth; this fiction was revered by

---

11 I-Hsing is the now defunct Wade-Giles spelling of Mandarin Chinese. This has been replaced by the pinyin pronunciation and he is termed Yi Xing.

the bureaucrats for its ease and reviled by the scientists for its fiction. It therefore persisted into relatively modern times.

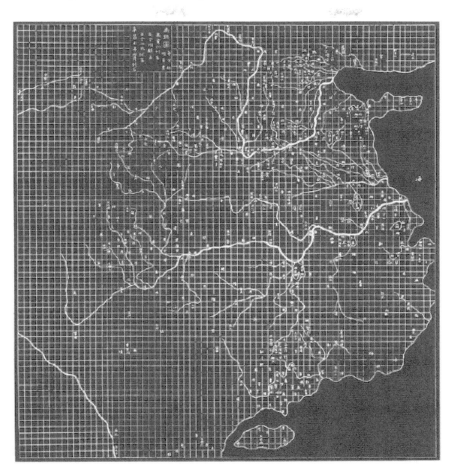

Yujitu – *Map of the Tracks of Yu*

It is not proposed to explore this subject in detail here; it is adequately dealt with in the article Chinese Metrology volume 2 of this series. However, certain points have been touched upon that make these Chinese observations pertinent to the history of the British-Gallic geography. Inasmuch the evidence suggests that geodesy and accurate map making has its roots deep in prehistory and in more recent times and later dynasties there appears to have been a falling away of the knowledge – however much one is conditioned by modern conceit to believe that the converse is true.

It was the initial observation, regarding Yi Xing's examination, that the origin of the survey in Vietnam, and the destination close to Lake Baikal, were two points on a due north meridian. Sometime later when enquiring into the history of the Yu Ji Tu, (a twelfth century map of China that is of the most startling accuracy) it was noted that many aspects of the map showed direct relationships with the survey of Yi Xing. The map is engraved on stone and was set up outside an institution of learning in Xi'an in 1137 AD; this was the oldest of the ancient capital cities of China; through the centuries it has been the capital of the Zhou, Qin, Han, Sui and Tang. The map was described by Joseph Needham as "the most remarkable cartographic work of its age in any culture."

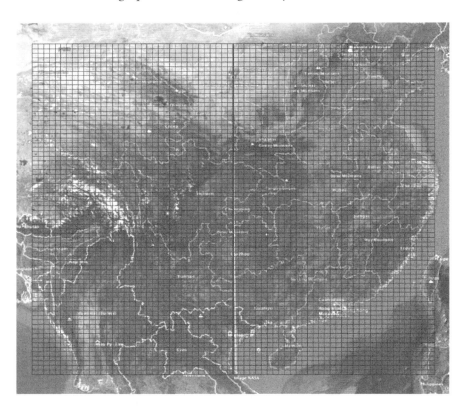

The grid of the Yujitu overlaying the physical map, Yi Xing's survey line is central.

The map is divided longitudinally and latitudinally into a series of squares, each square is 100 by 100 li. In this case the li in question would be the "short Tang li" which is 1,500 Tang feet (exactly equivalent to two Greek 600ft stadia). This is the first coincidence, because this would have been the module, (alongside the "long Tang li" of 1,800 Tang feet) that were used by the Tang dynasty monk Yi Xing, some 400 years before this map was installed at Xi'an. Note the accuracy of the Yangtse, Yellow and Mekong (to the south west) river courses, to the geographic reality.

The map extends over 24° from 17° to 41° degrees latitude; there are 73 east-west divisions to map and 70 north-south. The second coincidence is that the southern border of the map upon 17° is the latitude of Yi Xing's first survey station. The third coincidence, a big one, is the fact that Yi Xing's due north line of progression is exactly on the central line of the map. This is illustrated above= by a facsimile of the Yujitu grid being overlaid onto a Google earth map of China. This may all seem digressional to the Measures of Gaul but the distinct parallels between the inherited geographic knowledge of both Europe and of Asia are underscored through an appraisal of the history of the Yujitu. Although the stone stele on which the map is engraved was erected in 1137 AD, listed among the hundreds of place names engraved thereon were a great many fiefs and prefectures that had been defunct, in some cases, for centuries.[12] Due to this fact, it is generally believed that the map is a copy from a much older source. The consensus of opinion has it that the original, if not drawn, had been adequately described by Pei Xiu, minister, geographer and cartographer of the third century Three Kingdoms and Jin dynasty.

Probing further, the maps likely original inspiration is pushed back further back to Zhang Heng, a first century Han polymath, accredited with being the first to propose map-making on a grid coordinate system; it was he that had enthused Pei Xiu. Thereafter, as in Europe, the facts become sketchy and information retreats into the realm of the mythic and legendary; in Europe we have accreditation as to cartographic knowledge ascribed to the shadowy – king Belinus, or Herakles, roads dedicated by the Celts to such as Rhiannon (Christianised as Ellen and Brunhilde), but

---

12   Akin, Alexander and Mumford, David, *"Yu Laid Out the Lands"* geo referencing the Chinese *Yujitu. Cartography and Geographic Information Science* vol. 39, No. 3 July 2012

nothing solidly evidential. In China, there is much more than this mere veneer regarding the origins of map making because the legends are far more specific.

The very name – Yu Ji Tu – meaning "Map of the Tracks of Yu" is a direct reference to the mighty works of engineering carried out by the legendary deity Yu the Great; eighth great-grandson of the founder of China – The Yellow Emperor. The map was believed to have its origins as early as the twenty first century BC as his working diagram for the flood control works that he carried out, in order to tame the worst of the depredations wrought by the repetitive flooding of the Yangtze and Yellow Rivers.

It cannot be denied that throughout the world, works requiring an advanced cartographic knowledge, for flood control in Asia and Egypt and the building of a sophisticated road system throughout Europe, had been practiced in antiquity by a means now unknown. Although little is known of the methods by which this was accomplished, at least the undeniable evidence is there in what remains. A major tool of the ancient engineers must have been an advanced system of metrology. Unfortunately, this is largely abstract, and due to this, it attracts but little attention. Metrology is the most supreme of our ancestral inheritances, although its origins are lost far into prehistory, it survives and is, though fragmentary, still in use as an entirely practical system. It is the one ancient science at the greatest risk of disappearing in its entirety – and quite deliberately at that, under the onslaught of the modernism of metrication.

Gallic reaping machine as described by Pliny.

Book VI

GREEK
STADIA
LENGTHS

PREAMBLE

Written in 2007, this was a simple and enjoyable article to shape up. Once again it concerns a series of constructions of the same nature and purpose – a cluster. A Greek stadium, the length of the footrace, was 600 feet. Because the stadia of Greece are of different lengths they are therefore ideal to illustrate the different lengths of the feet in common use. On the majority of the surviving tracks the starting blocks and finish lines are set in stone, there are also hundred foot markers beside many of the tracks. This enables good accuracy in module identification. The levels of correspondence to the megalithic and iron age foot lengths that have been pinpointed in the previous sections, is quite remakable.

Because these articles were not originally written to be presented altogether, the explanations of the basic structure may seem tiresomely repetitive; gloss over if necessary.

GREEK STADIA LENGTHS

6.1 OVERVIEW

The stadia mentioned in the title of this work are not the principle subject; they have been selected as an excellent example of the values of length exemplified in ancient metrology. Firstly, the lengths of the running tracks of ancient Greece from whence comes the term "stadium" will be investigated. Then other recorded values of the "stade" that are not associated with the race tracks will be specified and their metrology explained accordingly. Corroboration of the foot measures will be given in the appendices.

The reason that the English foot is used in this exercise is that this is the medium through which the subject may be best understood – the English foot itself being an ancient measurement that is related to all others. Metrology would never have been deciphered by means of the metric system. The adoption of the metre is the principle reason that the subject has remained veiled because the rational *numbers* that the old designers were expressing are destroyed. These relevant numbers are most easily detected from the point of view of the English foot because it is close to the median of all the possible feet, the smallest (Assyrian) at nine-tenths of the English foot and the longest (Russian) at one and one-sixth of the foot. First, a few words on general metrology.

It is acknowledged that there was a spread of values for the so called Greek foot and it has often been remarked that the English foot more closely resembles the Greek than the Roman. There is a very good reason for this in as much that the English foot is one of the series of the Greek feet. It was Flinders Petrie who first proposed that measures regularly varied by the 450th and 170th part. It is now possible to refine these figures to be the absolutes of the 440th and the 175th part, and if one thinks in terms of the English foot and places this in the Root position the essential spread of the measures may be portrayed as so:

GREEK/ ENGLISH	Root Reciprocal	Root	Root Canonical	Root Geographic
	0.994318ft	1ft	1.005714ft	1.01146ft
	(30.307cm)	(30.479cm)	(30.654cm)	(30.829cm)

	Standard Reciprocal	Standard	Standard Canonical	Standard Geog.
	0.996578ft	1.00227ft2	1.008ft	1.01376ft
	(30.376cm)	(30.549cm)	(30.724cm)	(30.899cm)

The first line is appended Root, the second Standard. These modules increase by the 175th part horizontally and the 440th part vertically, they increase from the English foot to achieve acknowledged values of the Greek foot. The Parthenon foot, for example, is very clearly given as 1.01376 which is the 440th part greater than the Attic 1.0114612ft that will be demonstrated, particularly at Corinth. Because we are viewing the base measurement as the English foot, the above table is also the formula numbers by which all other modules may be classified.

It is easily observable how the English foot may be used as the identification code for other modules, the Root Geographic value at 1.01146122 is plainly the *number* $(176/175)^2$ Viewed as a number it is the formula, and as we have seen, once any module is seen to have a direct single fractional link with a Root value it will be subject to these multiplications to maintain the same fractional link with other values. In other words, all of the modules used in the ancient world are universally subject to these fractional variations. Having this knowledge at ones disposal enables one to simplify metrological analysis of the monuments.

These are the race tracks that are sufficiently intact to supply adequate data for their intended length:

Race Tracks	Measured	Calculated	Variant classification		
Halieis	166.5	166.478 m	600ft of .910315ft	Root Geog. Assyrian	
		Alternatively	500ft of 1.092378ft	Root Geographic Belgic	
Delphi	177.55 m	177.578 m	600ft of .97101ft	Root Geographic Roman	
Nemea	178	177.98 m	600ft of .9732096ft	Standard Geog. Roman	
Epidaurus	181.30 m	181.209 m	600ft of .99082ft	Root Geog. cmn Egyptian	
Isthmia (2)	181.20 m	181.209 m	600ft of .99082ft	Root Geog. cmn Egyptian	

Isthmia (1)	192.24 m	192.46 m	600ft of 1.052386ft	Standard Persian
Priene	191.39 m	See text Priene		
Miletus	192.27 m	192.4602 m	600ft of 1.052386ft	Standard Persian
Olympia	192.28 m	192.4602 m	600ft of 1.052386ft	Standard Persian

The above terminology concerning the variants will be explained as we progress. The most obvious thing that is apparent is that a whole range of measures were used in Greece and that there were variations in each of the individual modules. There is nothing arbitrary in the systems, they have absolute values. This phenomenon is not peculiar to Greece, it was common practice in all nations and they all used the same measures. Although these observed variations are lately attributed to slackness, there are perfectly good reasons for them.

6.2 HALIEIS

Due to land subsidence this track is now submerged in six feet of water, however the starting lines are in situ and have been accurately measured. This stadium is a very good example of the nature of metrology because it has an interpretation of being either the traditional 600ft length of the running track or the 500ft length of an itinerary stade (1/10th of a mile). The length is given by David Gilman Romano as 166.5 metres and is calculated as 546.19ft (166.478m). When analysing a measurement one simply has to reduce the number to its constituent foot, in this case, because it is closest to 500 in round figures then one divides by that number. The resultant number, in this case 1.092378ft, is then immediately identifiable and classifiable.

Aerial photograph showing the outline of the submerged running track

This foot is widely known as the Drusian or Belgic foot at this precise value. As this foot is nine eighths of the Roman foot of .971ft it was adopted by the governor of Gaul, Nero Claudius Drusus, as the standard for Gallic and Roman exchange. Another variant of the Belgic foot is 1.08ft and one may term this, as a means of classification, the *Root* value. So, when one is confronted by a number such as 1.092378, you may divide it by its Root value and the resultant number, in this case 1.0114612, can be classified as a *Root Geographic Belgic foot*. This will be further explained in the following Attic stadium commentary.

It is the alternative measurement, that of .910315ft, that is the probable intention of the designer in order for the race track to conform to the 600ft formula. Often referred to as the lesser or Italic foot it should be divided by the rational .9 in order to classify it. .9ft is the value identified as the Assyrian foot by Oppert from the ruins of Mesopotamian Khorsabad. .910315 divided by the Root .9 again equals 1.0114612 it is therefore classified as a *Root Geographic* Assyrian foot. Plainly the Assyrian foot has the association with the Belgic as 5 to 6 and relationships of such unit fractions demonstrably link all of the systems of antiquity.

A passing note that is relevant from the *Bryn Mawr Classical Review* 95.09.19 by David W.J. Gill, University of Wales Swansea:

*The length used at Halieis is surprising given that the layout of the city itself seems to have used a foot of c. 0.313 m (*T.D. Boyd and M.H. Jameson, Hesperia 50 [1981] 332*). However this may indicate one of two things: either that the stadion was laid out at a different time from the rest of the city, or that the running track was not a stadion in length.*

The point here being that there is absolutely no necessity that the foot values should be the same, one very often finds different feet used in a single construction.

Incidentally, the value for the city layout as mentioned by David Gill would more properly be a foot, not of 313mm but 313.5mm which is the Root common Greek foot of 1.028571ft as a fraction of the English it is 36/35.

6.3 DELPHI

Delphi Stadium. The foreground figures are standing on the starting line,
shown right.

The ancient stadium is situated high up the hill, beyond the sacred way and the Theatre. It was built in the 5th century BC and it was remodelled several times during the centuries. Its present form was acquired in the 2nd century AD when Herodus Atticus financed the stone seating and the arched entrance. Its stone seats could seat around 6,500 spectators, and it was used extensively during the Pythian and Panhellenic games for athletic events and for music festivals. Its track is 177.55 m long (582.5ft.), and 25.50 m wide.

With a measured length of 177.55 metres this corresponds to the theoretical length of 177.5768 metres to within an inch. This is the value of 600 Root Geographic Roman feet (Attic) .971ft and as well as being 440 to 441 of the following Nemean stadion this track has a relationship to the length of the Halieis stadion as sixteen to fifteen.

6.4 NEMEA

*Sheltered Nemea, partially built into the side of
the hill the slope of which was adapted for seating*

The Stadium, which could accommodate 40,000 spectators, was built 400m SE of the Temple of Zeus The track (total length of 178m) was bordered by a stone water-channel with stone basins at intervals for drinking water. The stone starting line was on its western extremity.

Stone drinking and wash basins for competitors

The Panhellenic games were held every two years in honour of Opheltes and it was built at the end of the 4th century B.C. In about 270 B.C., the games were transferred to Argos. Despite an attempt in 235 B.C. of Aratos of Sicyon to bring them back to Nemea (and indeed, for a while, they took place alternately in Nemea and Argos), the games were permanently transferred to Argos not long after. Unfortunately the length of the Argos stadion is unknown.

This stadium was excavated in 1974-1981 by the American School of Classical Studies (University of Berkeley, California) under the direction of S. Miller.

As with the stadion at Delphi this track is in terms of the Roman foot (confusingly but more correctly called Attic). It is 600ft of .9732096ft and is 441 to 440 of the Root Geographic foot of Delphi. This is termed a Standard Geographic Roman foot. The accuracy here is quite phenomenal, measured as 178 metres its calculated length is 177.98m, this is within ¾ of an inch between the measured and theoretical values.

This foot has a wide distribution; it was identified by Petrie as the Pelasgo foot and later as the Etrurian foot. One hundred of them are precisely the inner diameter of the Stonehenge sarsen circle. *Petrie, Stonehenge: Plans, Description, and Theories.* Although the difference between this Standard Geographic and the Root Geographic of the Delphi track is only the 440th part, at the distance of 600ft, this would amount to the distinctly measurable 40 centimetres.

6.5 EPIDAURUS

Epidaurus, western Peloponnese on the Saronic Gulf

"P. Kavvadias spent much of his life excavating Epidaurus, beginning in 1881. The French School took part early in the 20th century, followed by archaeologists J. Papadhimitriou and V. Lambrinoudhakis. The stadium was built between two small hills, and is 196.44 meters long, 23 m. wide, and has a 181.30 m. long running track. There were 22 rows of seats on the north side (right) and 14 rows on the south."

The overall arena length at 196.44 metres has a dual interpretation as has the running track at Halieis. Because of the fractional integration of metrology this length may be viewed as a 600ft stade of the so-called Doric foot of 1.07386 (Standard classification), or a 625ft (1/8th of a mile) furlong of the common Greek foot of 1.030909ft also at Standard. This simply means that the Doric and common Greek feet are related by the ratio of 25 to 24. (A similar interpretation to the overall length at Athens).

The length of the race track is given as 181.3 metres and the theoretical length is 181.209m. This is 600 common Egyptian feet of the Root Geographic classification, accurate with the measured length to about 3 ½ inches. Common Egyptian feet were widely used, certainly throughout the classical world and demonstrably in Britain at an even earlier period.

6.6 ISTHMIA (1)

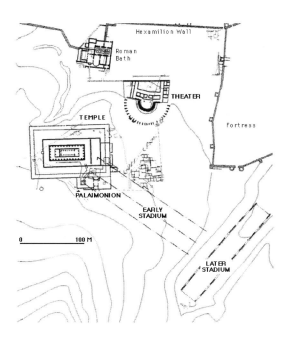

Isthmia, also on the Saronic Gulf, named from the Isthmus of Corinth

The later race track, Palaimonion, at Isthmia was built to replace the earlier, 6th century BC track, shown on the right. It was constructed 250 metres to the south east of the original longer track at about the time of Alexander. It is being dealt with first because it is directly related to the Epidaurus stadium by also being 600 common Egyptian feet. The measured length is given as 181.20m and the calculated value is 181.209m. Thus this track is even closer to the calculated value than Epidaurus; indeed it is less than ½ an inch shorter. The constituent foot is 302mm, or .99082ft, (Root Geographic) it is the common Egyptian, related as six to seven of the royal Egyptian other examples of its use in Greece will be given. The use of this particular foot is confirmed by the width of the sixteen lanes each being exactly five of these feet or 1.51m. *(Regarding this foot see "heptastadium" appendix, p.244).*

6.7 ISTHMIA (2)

The second race track was about 36 feet or 11 metres shorter than this original whose estimated length was computed from the one preserved end of the track and is given as 192.24m. For this is calculated as intended 192.46m and over the 600 Standard Persian feet (631.43ft) amounts to less than 9 inches discrepancy. There is no direct metrological link between the two track lengths of Isthmia one and two.

Dimensional information: International Olympic Academy twentieth session June 1980 Ancient Olympia 70.94-99 The Isthmian Games by Prof. Oscar Broneer (USA), Director, American School of Archaeology).

Also Deiter Lelgemann in : Recovery of the Ancient System of Foot/ Cubit/Stadion – Length Unit gives the value of this stadion as ideally 192.42m.

Fragment of the Isthmia stadium where the Athletes would be seated and where the hysplex or starting gate mechanism would be controlled from.

6.8 Miletus

The picture is a reconstruction.

The original Miletus race track is virtually the same as Isthmia, at measured 192.27m it varies from the calculated 192.46m by about 7 ½ inches, the same as the previous two tracks of Epidaurus and Isthmia is *calculated* as 192.46m. The width is given as 29.56m is therefore within the range of 100 Roman feet. 192.4602 m is 600 Standard Persian feet of 1.052386ft. *"The fact that the stadium conforms to the grid plan of Miletus has led some scholars to conclude that when the city was newly laid out in ca. 479 B.C., space was already allocated for the stadium. The fact that the stadium was not constructed until the second century B.C., however, is clear from its building inscription, architectural details, and relationship to the gymnasium to the west. The stadium lacks the curved ends or "sphendone" typical of stadia of the Roman period, and is similar to the ground plans of the stadia at Olympia, Epidaurus and Priene. Another similarity between the stadium at Miletus and the Stadium at Priene is the form and arrangement of the starting blocks or aphesis, although their exact mechanism remains unclear."* **Sarah Cormack, Perseus Dictionary, Tufts.**

The above extract was included because the statement that the Miletus city ground plan conforms to the stadium is quite unusual. The majority of city blocks, particularly in the Hippodamaeon system, were in increments of the 120ft Attic actus which is identical to an Egyptian plethron (two thirds of their "khet"). This convention was universally continued by the Romans at the identical lengths. (*See appendix, heptastadion, p. 244*).

6.9 OLYMPIA

Olympia

192.4602 m = 600 Standard Persian feet of 1.052386ft This is the fourth running track of these dimensions and is not materially different to the value given by Dieter Lelgemann in his *Recovery of the Ancient System of Foot/Cubit/Stadion – Length Unit.* The fact that so many commentators find all of these four stadia to be exactly 192.27m immediately arouses suspicion, for the simple reason that persons measuring the *same* object will reach slightly different lengths. One therefore suspects that there is a certain amount of copyist information in the statements.

Olympia is the site of the original Olympic Games which are virtually impossible to date. The Greeks reckoned historical time in four-year Olympiads starting from the date 772 to 776 BC. However, Iphitos is reputed to have revived the games from a far earlier age. He was a descendant of the mythical Oxylos, an ancient sovereign of Elis. When consulting the Oracle of Delphi in order to rid his land of warfare and disease, he was instructed to revive the games for the well being of the nation. As historians are generally agreed that the historical Iphitos dates from some time in the ninth century BC, *(Pausanias 5.4.5-6)* then the original games must have originated long ages before.

Hercules is also credited with modifying the games. One legend has it that he marked out the length of the track by walking heel to toe 600 times. As the anatomical foot is one seventh of the height, this would make Hercules 7ft 4½ inches tall (2.25m). Another legend has it that the stadion was the distance Hercules could run in a single breath. Additionally, 220 yards is a 600 Saxon feet stadion and survived as the length of a foot race amongst the British and their colonies till modern times. Alas now abolished by creeping metrication and replaced by the 200 metres, (which is 600 Belgic feet).

6.10 PRIENE

The Stadium of Priene is given as 191.39m and this length must be dismissed as its 600th part conforms to no known foot measure. A more accurate estimate of the original length should be sought before comment can be made. As the large part of this stadium has slipped down the hillside on which it was built, this definition has obvious difficulties. At 627.92ft its conjectured foot at 1.04653ft is difficult to categorise, it may be of the lesser Persian classifications or one of the greater common Greek feet.

Priene showing its ruinous condition

The difficulties are apparent from the picture. The starting blocks are in the foreground, the scattered remnants of the seating masonry were originally built into the slope on the left; much of the track has disappeared to the right.

The plan of the ancient town showing the race track at the bottom of the picture with the gymnasium building to its left; it illustrates the cliff that it was constructed upon. It was the erosion of the cliff that produced the ruin that exists today.

6.11 ATHENS

Athens sphendone built for the Olympic Games of 1896

The original stadium at Athens was built at the expense of Lykourgos in 330 BC. In 144 AD Herodes Atticus restored the Stadium, giving it the form that was found by Ernst Ziller at the 1870 excavation: the horseshoe construction with an arena 204.07 meters long. The present restoration of the Stadium was conducted by G. Averof for the staging of the revived Olympics from 1896.

The Athens stadium as it survived from antiquity

It would not be possible to propose a concrete value for the length of this course as we are able to from the survival of the starting blocks at other stadia. It is sufficient to remark that the overall length of 204.07 metres is very close, at 204.1m, to a value of a furlong of 625 feet of 1071428ft which were termed "Doric" by Dinsmoor. (A more correct terminology would be "lesser Belgic").

However, a value of 185 metres is widely accepted as an Attic 600 feet stadium; this may now be exactly expressed as 184.976 metres because this would be exactly 600 feet of 1.0114612ft or (176/175) squared in terms of the English foot. This value is indubitably identified at Corinth.

6.12 CORINTH

Based upon:
Finding the Center of a Circular Starting Line in an Ancient Greek Stadium - by Chris Rorres and David Gilman Romano

Fragments of a circular running track were found here and fortunately included the starting line where the athlete's standing positions were clearly marked. The track was circular, and as the standing positions were in single file it is conjectured by the mathematicians, Chris Rorres and David Gilman Romano, that it was primarily used for long distance races.

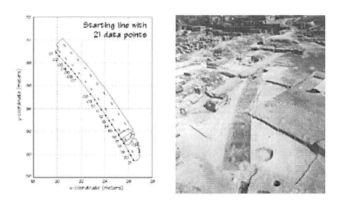

The starting line with data points

The archaeologists who had preceded them had calculated the radius of the circle which the fragmentary circumference gave them the data points. There are 21 data points remaining on the circumference i.e. the athlete's standing points, and taking a total of 680 triplets from combinations of the points (three points circles method) led them to believe that the radius was 56.242 meters. Rorres and Romano using more elaborate and accurate methods on the data, that of least-squares circle, concluded that the radius was 53.96 metres.

Another conclusion that they reached was that the starting positions had been laid at increments of roughly one degree upon the perimeter. In their own words:

"The average angle between the twelve starting positions is 1:019 degrees. Although the range of the angles is rather large, this average angle is sufficiently close to one degree to warrant attention. Among the Greeks, the earliest known use of the degree as a unit of angular measurement is found in the writings of Hypsicles in the second half of the second century B.C. He adopted the Babylonian practice of dividing the twelve signs of the zodiac into 30 equal parts. The choice of 30 is probably because the sun takes about 30 days to pass through each sign of the zodiac. That is to say, the division of a circle into 360 parts can probably be traced to the fact that there are roughly 360 days in a year. It may be that in laying out the starting positions, the Corinthians used the angle travelled by the sun along the zodiac in one day as the angle allotted each runner. If so, the Corinthian starting line is evidence of an early use of the degree as a unit of angular measurement among the Greeks."

The explanation is far more prosaic – and exact – than that. In order to see what is happening here from a metrological standpoint is to first convert to feet. Assuming that the data produced by the mathematicians is more accurate than that of the archaeologists, then their estimate of 53.96m for the radius is 177.03427ft and *invariably* ancient radii may be divided by seven to find the module. One seventh is 25.2906ft and this is therefore divided by the obvious 25 as the nearest whole number, it then produces the constituent foot module of 1.0116244ft. This is then the Root Geographic Greek foot to accuracy of one part in 6198 or 99.984%, (one third of an inch in 54 metres). The radius in terms of the module is therefore seven five feet paces.

This radius yields a perimeter of 1100 Greek feet of 1.0114612ft which is 440 of the five feet paces detected in the radius. 22/7 was invariably the pi ratio unless the diameter was a multiple of 4, they would then use 25/8 and the longer by 175th part module in the perimeter. 22/7 = 176 and 25/8 = 175, this essentially means that they were still using 22/7.

The obvious conclusion is that the data points on the starting line are spaced exactly three Greek feet apart and this gives not quite the 1.019 degrees but 1.0185185 degrees. This means that there was no intention that the perimeter should be divided by 360 or 365, it simply means that 3 feet go into 1100 feet 366.666 times. The spacing between the standing points of 3.03438 feet is recognisable as a 3 feet module but .92488 metres is not. This is the problem with metrology and the metre, the metre has no sensible sub divisions that resemble the feet that are the basis of ancient metrological design, as a consequence the essential *numbers* are lost. However, Rorres and Romano have made a most skilful reconstruction of an ancient module, even though they did not recognise it.

The proof of the system as outlined is the fact that measures from the field so often conform to extreme degrees of accuracy with those that are proposed by the simple arithmetic of this general theory of ancient metrology.

APPENDIX I

Notes on integrated metrology and stadia in general .

The Alexandrian island of Pharos was connected to the mainland by a mole that was seven stadia in length and this was called the Heptastadium. As the length is given as 1,260 metres then this length divided by seven should give a fairly accurate value of one of the stadia lengths accepted in the ancient world. Seven stadia are 4,200 feet and the nearest value to this length is the Root Canonical common Egyptian foot of .9852ft, 4,200 of which are 1261 metres. (It must be admitted that accurate estimates of this distance are difficult to come by, but the above value is the most often quoted. Recent geophysical exploration of this causeway may produce fully reliable data).

Pharos and its connecting causeway

We have encountered the common Egyptian foot at both Epidaurus and Isthmia stadia as 600 feet of .99082ft. This is the Root Geographic value which is 176 to 175 of the Alexandrian heptastadium based on .9852ft. It is generally believed that the term Heptastadium refers exclusively to the connecting causeway of Pharos to the mainland, but the term is encountered in the writings of Heron of Byzantium *(De Mensura p368)*. (this would therefore have to have been before the tenth century).

He gives the value of a mile which is 5400 Italian feet and states that this is also 4500 feet of the royal or Phileaterian foot, also that the royal foot is 16 finger breadths and the Italian 13½, clearly this is the Roman foot compared to the royal Egyptian. He states that the Royal foot is seven stadia to the mile and 7 × 600 royal Egyptian feet of 1.14285ft = 4800ft and this is exactly the Roman mile of 5,000ft of .96ft.

Much speculation has been forwarded about the solutions to Heron's statements, although it is obvious that a mile of 4,500 royal Egyptian feet is exactly 5,000 common Greek feet — mille passum. Furthermore, the connection to make 13½ royal digits equal to 16 Roman digits requires a classification change of the 175th part to the Roman foot, from Root to Root Canonical for example — if that is the variant one is discussing. Complicated? No. Complex? Yes.

Appendix 11

Here one of the feet will be considered to explore its pedigree to illuminate the reasoning. The obvious module to choose is that which is the most prevalent in the stadia lengths and this is the foot of Olympia. It is also the foot of Miletus, Isthmia 1 and it is strongly suspected that it, or one of its variants, is the intended foot of Priene.

Isthmia (1) 192.24 m, or 600ft of 1.052386ft Standard Persian
Miletus 192.27 m, or 600ft of 1.052386ft Standard Persian
Olympia 192.28 m, or 600ft of 1.052386ft Standard Persian
Priene 191.39 m. See text **Priene**

As stated, this foot has a Persian root. The value of the Royal Persian Cubit of Darius the Great is given as 2.1ft, it is therefore a two feet cubit or "*dupondium*" and the constituent foot is 1.05ft, or 21 to 20 English feet. It is one "Root" value of the variations because of this *unit fraction* relationship; however reduced by 1.008 it is 1.04166ft that is the *lesser Root* because this is also the unit fraction 25 to 24 English feet. All of the variants would therefore be termed Persian, from wherever they be found, appended by the classification terminology. Here are listed all of the core variations of this Persian foot:

PERSIAN	Root Reciprocal	Root	Root Canonical	Root Geographic
	1.044034ft	1.05ft	1.056ft	1.062304ft
	(31.818cm)	(32.004cm)	(32.183cm)	(32.367cm)

	Standard Reciprocal	Standard	Standard Canonical	Standard Geog.
	1.046407ft	1.052386ft	1.0584ft	1.064448ft
	(31.890cm)	(32.073cm)	(32.256cm)	(32.440cm)

All of the values are dictated by the fractions that occur in the original table on page 226.

Closely given values to these listed are found in the following: Petrie recorded three values of this two feet Persian measure

At Abydos Egypt	2.09281ft	= foot of	1.046405
In Palestine:	2.1066ft	= foot of	1.0533
In Persia:	2.1116ft	= foot of	1.058

Greaves recorded the lesser pike in Turkey:

	2.13125ft	= foot of	1.06562

Arabic Hashimi cubit

	2.12889ft	= foot of	1.064448
English mile of	5280 ft	= 5000ft of	1.056

The above Hashimi value was adopted by Charlemagne from Harun al Raschid as the original pied de roi. A much older standard of this foot being used in France is identified by Jacques Dassié from analysis of Gallic leagues as 1.056ft. This league of 7500 feet is exactly 1 ½ English miles and the Arabian parasang is exactly three English miles. The parasang is 30 stadia therefore they are 500 feet itinerary stadia of 1.056ft.

Quite the most unequivocal example of the use of this Persian foot in Greece is the metrological relief of Salamis. Found in a church, it had been salvaged from an ancient building, this artefact was brought to general attention at the 13th Congress of Classical Archaeology in 1988 by Ifigenia Dekoulakou-Sideris.

The Salamis relief

It is a depiction of different standards of length that were most likely used as a reference in the locality. The drawing above faithfully records her findings, the foot rule is 322mm, the span is 242mm, the footprint is 301mm and the cubit above the arm is 487mm.

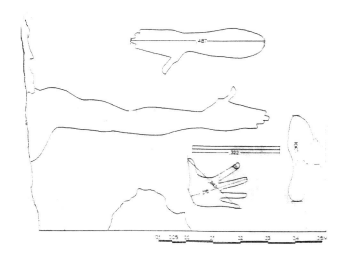

Values according to Sideris 1998

The modules conform very correctly to lengths established elsewhere and agree with the lengths proposed in the tables with a high degree of accuracy. The foot rule is the basis of the span; the span is the half cubit that would be comprised of 1½ times the foot rule. It is the Root Canonical Persian foot, 1.056ft, to within a tenth of a millimetre.

The cubit above the arm however is the Standard Geographic value (1.008 times the foot rule classification at 1.064448ft × 1.5). The cubit is one third of a millimetre different to its intended value of 486.66 mm, measured as 487mm.

With even greater accuracy to the preceding, the foot of 301mm is the common Egyptian foot of the Standard Canonical classification, which has an absolute value of 300.97mm. It is the common Egyptian foot that was used in the layout of Epidaurus and at Isthmia at the Root Geographic classification which is slightly longer. These values are extremely close together, but they would be distinguishable over the 600 feet track by a difference exceeding 2 feet.

These "Persian" feet are in no way Persian, nor is the common Egyptian "Egyptian." These feet are found universally. They date from remote antiquity from a system of measures that is so tightly integrated that it could not have evolved. It is based upon pure number and how numbers behave to rationalise everything into integers. The modules are recognisable because they are invariably expressed into rational numbers when classified by the ancient artisans. For example, the Alexandrian causeway to Pharos is an arbitrary length that is rationalised into an integral module, which is why this particular module was selected.

One encounters the same selection process on the arbitrary length of the Siloam tunnel. From the Gihon spring a tunnel was dug through the solid rock of the hill of Ophel which resulted in the reservoir known as the pool of Siloam inside the lower city of Jerusalem. This was an arbitrary distance but is stated in a contemporary inscription to be 1200 cubits:

This is the record of how the tunnel was breached. While [the excavators were wielding] their pickaxes, each man toward his co-worker, and while there were yet three cubits for the brea[ch,] a voice [was hea]rd each man calling to his co-worker; because there was a cavity in the rock (extending) from the south to [the north]. So on the day of the breach, the excavators struck, each man to meet his co-worker, pick-axe against pick-axe. Then the water flowed from the spring to the pool, a distance of one thousand and two hundred cubits. One hundred cubits was the height of the rock above the heads of the excavat[ors].

Siloam tunnel inscription of Hezekiah, translation above (c. 650 BC)

The most reliable source for this distance is that of *C. R. Conder, in the "The Siloam Tunnel." Palestine Exploration Fund Quarterly Statement 14 (1882)*, he gives the length as 537.6 metres.

Part of the tunnel that has many twists and turns beneath Jerusalem

First convert to feet, it is 1,763.78ft, then reduce this to its constituent foot by dividing by 1,800 which is .979878ft and this is seen to be almost exactly the Root common Egyptian foot which is .9795918ft (a difference of six inches in the whole length). This is the Root value because it is exactly 48 to 49 of the English foot; it is also 6 to 7 of the royal Egyptian foot which in turn is 8 to 7 of the English foot. At Epidaurus and Isthmia racetracks it is 1.0114612 times longer than Root (Root Geographic), on the Salamis relief it is 1.008 times longer than Root (Standard Canonical) and in the Heptastadium it is 1.0057142 times the Root (Root Canonical).

Hence it would seem that clear rational integers have to be expressed in the finished artefact, and this is how ancient metrology was applied. As stated, if one of these arbitrary lengths such as the length of the Siloam tunnel or the Heptastadion of Alexandria were under consideration then they would select a module to express the length in a rational number. This would be simple to apply because all of the possible modules were concurrently used in all nations.

The only reason that they have national or cultural terminologies Sumerian, Egyptian, Belgic, Roman etc or Doric, Attic, Corinthian etc, is that they were selected from a wide range for the purposes of bureaucratic regulation within those nations or cities. It is a system of elegance, beauty and wit and the English speaking nations should find it easy enough to grasp, at least for a few years until it is overcome by the stealth of metrication and finally extinguished.

John Neal, London, July 2006

INDEX